64th Israel Annual Conference on Aerospace Sciences (IACAS-2025)

Haifa, Israel
20 March 2025

ISBN: 979-8-3313-1612-9

Printed from e-media with permission by:

Curran Associates, Inc.
57 Morehouse Lane
Red Hook, NY 12571

Some format issues inherent in the e-media version may also appear in this print version.

Printed by Curran Associates, Inc. (2025)

Faculty of Aerospace Engineering
Technion - Israel Institute of Technology
Department of Aerospace Technology
Haifa 32000
Israel

Phone: (04) 829-2260
Fax: (04) 829-2030

aerospace.technion.ac.il/

Additional copies of this publication are available from:

Curran Associates, Inc.
57 Morehouse Lane
Red Hook, NY 12571 USA
Phone: 845-758-0400
Fax: 845-758-2633
Email: curran@proceedings.com
Web: www.proceedings.com

IACAS

The 64th Israel Annual Conference on Aerospace Science

Thursday 20 March, 2025

PROGRAM

Event	Time/Venue
• Opening • Plenary Lectures • Lunch • Exhibition stands of IACAS sponsors	9:00-13:30 **Churchill Hall, the Technion**
• Regular Sessions	13:30-18:00 **Faculty of Aerospace Engineering, Technion**

sponsors and supporters

TECHNION Israel Institute of Technology

IAI

RAFAEL ADVANCED DEFENSE SYSTEMS LTD.

Elbit Systems

Systematics

ISCFDC

National Hypersonics Center

DEPARTMENT OF AEROSPACE ENGINEERING

BMGC

ISAA

IACAS

IACAS-2025 Program - Table of Content

Thursday 20.3.2025		Technion, Haifa
Time/Code	Session	Hall
8:45-12:00		
9:00-9:10	Welcoming Address: Prof. Uri Sivan, Technion President	**Churchill Hall**
9:10-9:15	IACAS Opening: Dr. Judith Hocherman-Frommer, Chair IACAS-2025	
9:15-10:05	Colonel (Res.) Dror Ben-David	
	Coffee Break *10:05-10:30*	
10:30-11:20	Professor Wei Shyy, Chair Professor at the Hong Kong University of Science and Technology (Guangzhou)	
11:20-12:10	David A. Vallado, Senior Research Astrodynamicist, COMSPOC	
	Lunch *12:10-13:30*	
13:30-15:30		
ThL1T1	Guidance, Navigation, and Control	A
ThL1T2	Computational Fluid Dynamics	B
ThL1T3	Flow Stability and Control I	C
ThL1T4	Propulsion and Combustion	D
ThL1T5	Aerodynamics & Aeroacoustics	E
ThL1T6	Aeroelasticity & Fluid-Structure Interaction	F
ThL1T7	Astrodynamics and Space Systems I	G
ThL1T8	Aerospace Design	H
	Coffee Break *15:30-16:00*	
16:00-18:00		
ThL2T1	Propulsion, Combustion and Energy Systems	A
ThL2T2	Flow Stability and Control II	B
ThL2T3	Aerospace Systems and MDO	C
ThL2T4	AI and Autonomy	D
ThL2T5	Aerodynamics, Hydrodynamics and Aeroacoustics	E
ThL2T6	Astrodynamics and Space Systems II	F
ThL2T7	Solid Mechanics and Aerospace Materials	G

IACAS

Keynote Lectures

9:15-10:05 ThPl1

Colonel (Res.) Dror Ben-David

The seven trends that will shape our skies

Chair: Dr. Judith Hocherman-Frommer

About Colonel (Res.) Dror Ben-David

Colonel (Res.) Dror Ben-David served 27 years in the Israeli Air Force. commander of F16 and F15 squadrons. Head of the operational requirement department. B.A. in Physics with excellence, Technion. Currently, head of Neural Networks R&D Labs at Matrix IT, Israeli National Drone Initiative - Subject Matter Expert for operations. Autonomous Public Transportation SME. Civil Aviation - Cyber resilience SME, Israeli rep at ICAO.

Churchill Hall

Plenary Lectures

10:30-11:20 ThPl2

Professor Wei Shyy

Low altitude air mobility, low orbit satellites: opportunities, challenges and sustainability

Chair: Distinguished Professor Emeritus Daniel Weihs

In the present era of fast emergence and advancement of innovative ideas and new technologies, we are seeing unprecedented opportunities as well as challenges. In the civilian aerospace sector, low orbit satellites of altitude up to 2,000 km, equipped with advanced sensing and communication capabilities; and low altitude air mobility, including drones, un-crewed and crewed aerial vehicles, from ground to around 2 km, are two fast moving fronts. Individually and collaboratively, they present new frontiers in the global and regional contexts, including sustainability, commerce, emergency response and risk mitigation. In this lecture, examples will be presented along with technical issues to highlight implications and issues associated with these developments.

About Professor Wei Shyy

Professor Wei Shyy is a Chair Professor at the Hong Kong University of Science and Technology (Guangzhou). He was President of the Hong Kong University of Science and Technology and Chair Professor of Mechanical and Aerospace Engineering. Prior to joining HKUST in 2010, he was Clarence L. "Kelly" Johnson Collegiate Professor and Chairman of the Department of Aerospace Engineering of the University of Michigan.

He was the Principal Investigator of several multi-institutional research projects, funded by the US Government and industries, on future space transport, bio-inspired flight, and computational aero-science. He is General Editor of the Cambridge Aerospace Book Series published by the Cambridge University Press; and Co-Editor-in-Chief of Encyclopedia of Aerospace Engineering, a major reference work published by Wiley-Blackwell.

Professor Shyy is an elected member of Academia Sinica (Taipei), an agenda contributor of the World Economic Forum, and Fellow of the American Institute of Aeronautics and Astronautics (AIAA). He has received awards for his research and professional contributions, including the AIAA 2003 Pendray Aerospace Literature Award, the ASME 2005 Heat Transfer Memorial Award, and The Engineers' Council (Sherman Oaks, CA) 2009 Distinguished Educator Award. In 2021, the French Government made him an Officer of the Legion of Honor. In 2023, the International Conference on Computational & Experimental Engineering and Sciences (ICCES) honored him the Satya N. Atluri Award.

Plenary Lectures

11:20-12:10 ThPl3

Lt Col (USAF Ret.) David A. Vallado, Senior Research Astrodynamicist, COMSPOC

Challenges to Accurate LEO Satellite Prediction

Chair: Dr. Vladimir Martinusi

Accurate propagation of Low Earth Orbiting satellites is increasingly important for conjunction analyses as the region becomes more densely populated. Standard perturbation force models are successfully used for Orbit Determination to produce accurate epoch states. But what happens as these states are predicted into the future? The uncertainty increases rapidly, somewhat independent of the accuracy of the initial state. I review the dominant perturbing forces and focus on gravity and atmospheric drag as sources of the increased uncertainty in prediction.

About David A. Vallado

Lt Col (USAF Ret.) David A. Vallado is currently working as a Senior Research Astrodynamicist with COMSPOC in the Center for Space Standards and Innovation. He is the author of the advanced astrodynamics textbook, Fundamentals of Astrodynamics and Applications (5th edition, Microcosm, 2022). He is a Fellow in the American Astronautical Society (2006).

He attended the United States Air Force Academy and majored in Astronautical engineering, receiving his Bachelor of Science in 1980. Lt Col Vallado earned a Masters of Science in Systems Management from the University of Southern California in 1982. He then attended the Air Force Institute of Technology (AFIT), where he earned a Masters of Science in Astronautical Engineering in 1984.

His AF assignments include serving as a Project Officer for Stage I of the PEACEKEEPER missile, analysis for the Strategic Air Command Staff at the 544th Strategic Intelligence Wing, an instructor in the Department of Astronautics at the US Air Force Academy, several research scientist activities at AFRL, and analysis at US Space Command. After retiring from the Air Force, he was a principal engineer with Raytheon Intelligence and Information Systems in Denver CO and an Astrodynamics Researcher with Analytical Graphics Inc.

An avid mountain climber, he has hiked all 58 mountains in Colorado over 14000 feet, twice. Dave's other interests include jogging, biking, woodworking, house construction, classical piano, swimming, and square dancing.

IACAS

ThL1T1	Hall A

Guidance, Navigation and Control

Chair: Daniel Choukroun — Ben-Gurion University of the Negev
Co-Chair: Ilan Rusnak — Rafael

13:30-13:50 — **ThL1T1.1**

Ilan Rusnak — Rafael

13:50-14:10 — **ThL1T1.2**

Zamir Martinez — Technion
Daniel Zelazo — Technion

14:10-14:30 — **ThL1T1.3**

Ilan Rusnak — Rafael
Liat Peled-Eitan — Rafael

14:30-14:50 — **ThL1T1.4**

Sergey Nazarov — Technion
Per-Olof Gutman — Technion

14:50-15:10 — **ThL1T1.5**

Caitong Peng — Ben-Gurion University of the Negev
Daniel Choukroun — Ben-Gurion University of the Negev

15:10-15:30 — **ThL1T1.6**

Michael Urman — Technion
Oded Golan — Technion
Vitaly Shaferman — Technion

ThL1T2 Hall B

Computational Fluid Dynamics

Chair: Vassilis Theofilis — Technion

13:30-13:50 — ThL1T2.1

Ablation Computational Simulation using Ansys Fluent...30

Dvir Mendler — Ansys
Sunil Chadha — Ansys

13:50-14:10 — ThL1T2.2

Flow simulation of waveriders at off-design conditions...31

Angelos Klothakis — Technical University of Crete
Ioannis Nikolos — Technical University of Crete
Vassilis Theofilis — Technion

14:10-14:30 — ThL1T2.3

Wall Function LES using Fluent GPU solver and Fluent Rapid Octree mesh...34

Dvir Mendler — Ansys

14:30-14:50 — ThL1T2.4

Very large eddy simulations of the flow within a F-35's weapon bay with an internal store...35

Ron Efrati — Israeli Air Force
Hadar Ben-Gida — Israeli Air Force

14:50-15:10 — ThL1T2.5

Terrain Ridgelines Detection Based on Streamlines...36

Baruch E. Karlin

IACAS

ThL1T3	Hall C

Flow Stability and Control I

Chair: Michael Karp	Technion
Co-Chair: Alexandros Terzis	Technion

13:30-13:50	ThL1T3.1

Analytical flow solutions in rectangular microfluidic channels based on the darcy-brinkman model...39

Mario Del Mastro	Technion
Alexandros Terzis	Technion

13:50-14:10	ThL1T3.2

Stability analysis of the non-isentropic one-dimensional flow in a scramjet isolator...40

Shahaf Haiman	Technion
Michael Karp	Technion

14:10-14:30	ThL1T3.3

Linear stability of a weak viscous shock layer...41

Vassilis Theofilis	Technion

14:30-14:50	ThL1T3.4

Stability analysis of forced and unforced flow fields around cylinders in crossflow...43

Doron Schwartz	Technion
Michael Karp	Technion

14:50-15:10	ThL1T3.5

Particle clustering and turbulence modulation in eulerian particle-laden flow simulations...44

AJAY DHANKARGHARE	Technion
Yuval Dagan	Technion

15:10-15:30	ThL1T3.6

Investigation of oil flow topology, pressure distribution, and drag in separated flow over low-Reynolds-number airfoil...45

Roi Baruch	Technion
Igal Gluzman	Technion

ThL1T4	Hall D

Propulsion and Combustion

Chair: Dan Michaels — Technion
Co-Chair: Eran Sher — Technion

13:30-13:50	ThL1T4.1

Pressure effect on the burning rate of solid propellants with expandable graphite additives...46

Noa Edri — Technion
Alon Gany — Technion

13:50-14:10	ThL1T4.2

Investigating a gun-launched solid fuel ramjet projectile...49

Itamar Levitan — Technion
Alon Gany — Technion

14:10-14:30	ThL1T4.3

Enhancing water-breathing ramjet performance with boron additive...51

Sagi Dinisman — Technion
Alon Gany — Technion

14:30-14:50	ThL1T4.4

Experimental study of mode transition in a cavity-based dual-mode scramjet combustor...53

Alexander Dolnik — Technion
Dan Michaels — Technion

14:50-15:10	ThL1T4.5

Hydrogen propulsion: the fundamental equation for H2 in its liquid, saturated, and gaseous states...54

Eran Sher — Technion
Shir Levi — Technion

15:10-15:30	ThL1T4.6

Computer Vision Approach for Analysis of Numerical and Experimental Detonation Cellular Structure Images...55

Daniel Jalontzki — Tel Aviv University
Alon Zussman — Tel Aviv University
Guni Sharon — Texas A&M University
Sumedh Pendurkar — Texas A&M University
Yoram Kozak — Tel Aviv University

ThL1T5 Hall E

Aerodynamics & Aeroacoustics

Chair: Oksana Stalnov — Elbit Systems
Co-Chair: Hadar Ben-Gida — Technion

13:30-13:50 ThL1T5.1

Alexandros Peteinaris — Technion
Alexandros Terzis — Technion

13:50-14:10 ThL1T5.2

Elad Bar On — Technion
Yosef Pikel — Israeli Air Force
Hadar Ben-Gida — Israeli Air Force

14:10-14:30 ThL1T5.3

Sarah Sullivan — Rensselaer Polytechnic Institute
Evan Mahns — Rensselaer Polytechnic Institute
Miki Amitay — RPI

14:30-14:50 ThL1T5.4

Barak Deutscher — Technion
Oksana Stalnov — Elbit Systems
Hadar Ben-Gida — Technion

14:50-15:10 ThL1T5.5

Maayan Naschletashvili — Technion
Oksana Stalnov — Elbit Systems
Hadar Ben-Gida — Technion

15:10-15:30 ThL1T5.6

Dor Polonsky — Rafael
Gali Alon Tzezana — Rafael
Alon Dahan — Rafael

ThL1T6 Hall F

Aeroelasticity & Fluid-Structure Interaction

Chair: Maxim Freydin — Technion
Co-Chair: Moti Karpel — Technion

13:30-13:50 — **ThL1T6.1**

Static aeroelasticity and stability of very flexible swept wings...64

Bar Revivo — Technion
Daniella Raveh — Technion

13:50-14:10 — **ThL1T6.2**

Aeroelasticity of plates in supersonic channel flow...65

Maxim Freydin — Technion

14:10-14:30 — **ThL1T6.3**

Experimental investigation of the F-16 shock buffet with comparison to flight tests...67

Tzlil Nahom Jidovetski — Israeli Air Force
Michael Iovnovich — Israeli Air Force
Daniella Raveh — Technion

14:30-14:50 — **ThL1T6.4**

Aeroelastic coupling of hard maneuvering aircraft...72

Dor Naftaly — Israeli Air Force
Daniella Raveh — Technion

14:50-15:10 — **ThL1T6.5**

Numerical simulation of flutter flight tests using the parametric flutter margin method...74

Guy Gordon Biberstein — Technion
Moti Karpel — Technion

15:10-15:30 — **ThL1T6.6**

Advancements in in-flight aeroelastic sensing: real-time prediction of deflection, load, and rigid angle of attack using machine learning from ground test data...89

Ido Hauzer — Technion
Daniella Raveh — Technion

ThL1T7 Hall G

Astrodynamics and Space Systems I

Chair: Vladimir Martinusi Technion
Co-Chair: Alexander Batkhin Technion

13:30-13:50 ThL1T7.1

Network of symmetric periodic families in the Hill 3-body problem...91

Alexander Batkhin Technion

13:50-14:10 ThL1T7.2

Uncertainty propagation using Riemannian geometry...92

Vladimir Martinusi Technion

14:10-14:30 ThL1T7.3

Constant-magnitude low-thrust orbital transfer with final-approach targeting...95

Michal Pushkov Technion
SHRIBHARATH BALAKRISHNAN Technion
Pini Gurfil Technion

14:30-14:50 ThL1T7.4

Optimal control for a stationary solar sail around L1'...96

Naama Gilat Technion
Oded Golan Technion
Vitaly Shaferman Technion

14:50-15:10 ThL1T7.5

Technical requirements for optical sensors
in autonomous space situational awareness...97

Yuval Hammer Braude
Ido Ben Harosh Braude
Ivan Goncharov Braude
Ira Wolfson Braude
Elad Denenberg Braude Academic College of Engineering

15:10-15:30 ThL1T7.6

Satellite Signals-of-Opportunity-Based Navigation:
From Simulation to Real-World Implementation...110

Alex Frid Technion
Carlos Caravaca-Gallego Technion
Pini Gurfil Technion

ThL1T8	Hall H

Aerospace Design

Chair: Ilan Berlowitz	IBAero
Co-Chair: Ariel Dvorjetski	TBD

IACAS

ThL2T1	Hall A

Propulsion, Combustion and Energy Systems

Chair: Moshe Zilberman — Azrieli Academic College of Engineering
Co-Chair: David Yanuka — Technion

16:00-16:20	ThL2T1.1

Ariel Sharon — Technion
Yeshayahou Levy — Technion

16:20-16:40	ThL2T1.2

Moshe Zilberman — Azrieli Academic College of Engineering

16:40-17:00	ThL2T1.3

David Yanuka — Technion

17:00-17:20	ThL2T1.4

Elinor Kostjukovsky — Technion
Alon Gany — Technion

ThL2T2 Hall B

Flow Stability and Control II

Chair: Igal Gluzman — Technion
Co-Chair: Hadar Ben-Gida — Technion

16:00-16:20 — ThL2T2.1

Itamar Blumenfeld — Technion
Hadar Ben-Gida — Technion

16:20-16:40 — ThL2T2.2

Niv-Haim Mizrahi — Tel Aviv University
Elizaveta Dubrovskaya — Tel Aviv University
Ofek Katz — Tel Aviv University
Yarden Turgeman — Tel Aviv university
Alex Liberzon — Tel Aviv University
Oksana Stalnov — Elbit Systems

16:40-17:00 — ThL2T2.3

Vojtech Pezlar — Czech Technical University in Prague
Marlon Mathias — University of Sao Paulo
Vassilis Theofilis — Technion
Marcello Augusto Faraco Medeiros — University of Sao Paulo

17:00-17:20 — ThL2T2.4

Victor Barcelos Victorino — University of Sao Paulo
Felipe Oliveira Aguirre — University of Sao Paulo
Marcello Augusto Faraco Medeiros — University of Sao Paulo

17:20-17:40	ThL2T2.5

Felipe Oliveira Aguirre — University of Sao Paulo
PauloCelso Vieira Paino — University of Sao Paulo
Hadar Ben-Gida — Technion
Marcello Augusto Faraco Medeiros — University of Sao Paulo

17:40-18:00	ThL2T2.6

Ofek Frank-Shapir — Technion
Igal Gluzman — Technion

IACAS

ThL2T3 Hall C

Aerospace Systems and MDO

Chair: Yuval Freed — Israel Aerospace Industries
Co-Chair: Anna Clarke — Technion

16:00-16:20 ThL2T3.1

Boris Dorfman — Israel Aerospace Industries
Efrat Pinhas — Israel Aerospace Industries
Yuval Freed — Israel Aerospace Industries

16:20-16:40 ThL2T3.2

Yuval Freed — Israel Aerospace Industries

16:40-17:00 ThL2T3.3

Shaul Eliahou Niv — Israel Aerospace Industries
Yotam Gardosh — HUJI
Asaf Shiloah — HUJI

17:00-17:20 ThL2T3.4

Maayan Shimoni — Technion
Ian Jacobi — Technion
Anna Clarke — Technion

17:20-17:40 ThL2T3.5

Ido Braun — Technion
Joseph Z. Ben-Asher — Technion

ThL2T4 Hall D

AI and Autonomy

Chair: Itzik Klein — University of Haifa
Co-Chair: Vadim Indelman — Technion

16:00-16:20 ThL2T4.1

Tuvy Lemberg — Technion
Vadim Indelman — Technion

16:20-16:40 ThL2T4.2

Cristian Omat — Astronomical Institute of the Romanian Academy

16:40-17:00 ThL2T4.3

Barak Or — Technion

17:00-17:20 ThL2T4.4

Barak Or — Technion

17:20-17:40 ThL2T4.5

Tsoof Joels — Elbit Systems

17:40-18:00 ThL2T4.6

Itzik Klein — University of Haifa

ThL2T5 Hall E

Aerodynamics, Hydrodynamics and Aeroacoustics

Chair: Yuval Dagan — Technion
Co-Chair: Omry Magen — Tel Aviv University

16:00-16:20 ThL2T5.1

Boaz Ofarim — Technion
Orr Avni — Technion
Yuval Dagan — Technion

16:20-16:40 ThL2T5.2

Aharon Karon — Israel Aerospace Industries
Aleksandra Kvurt — Israel Aerospace Industries
Jenya Kazarin — Technion
Hadar Ben-Gida — Technion

16:40-17:00 ThL2T5.3

Elad Zur — Technion
Igal Gluzman — Technion

17:00-17:20 ThL2T5.4

Samruddhi Salunke — Technion
Igal Gluzman — Technion

17:20-17:40	ThL2T5.5

Rapid depressurization-induced flash boiling: a theoretical model for positive and negative pressure ranges...160

Omry Magen — Tel Aviv University
Yoram Kozak — Tel Aviv University
Laura DiLucchio — University of Pavia
Marco Marengo — University of Pavia
Tali Bar-Kohany — Tel Aviv University

17:40-18:00	ThL2T5.6

Eulerian sectional approach for particle erosion in compressible flows...162

Amir Loyevsky — Rafael
Ido Immer — Rafael
Yuval Dagan — Technion

ThL2T6	Hall F

Astrodynamics and Space Systems II

Chair: Vitaly Shaferman — Technion
Co-Chair: Moshe Golani — Technion

16:00-16:20	ThL2T6.1

Meir Nemirovsky — Technion
Pini Gurfil — Technion

16:20-16:40	ThL2T6.2

Yahli Drucker — Technion
Vitaly Shaferman — Technion

16:40-17:00	ThL2T6.3

Moshe Golani — Technion

17:00-17:20	ThL2T6.4

Azriel Lorber

17:20-17:40	ThL2T6.5

Revital Frenkel — Technion
Vitaly Shaferman — Technion

17:40-18:00	ThL2T6.6

Geffen Aharoni — Technion
Tamar Alperin — Technion
Naama Gilat — Technion
Iris Kanter — Technion
Almog Yanku — Technion
Alex Frid — Technion

ThL2T7	Hall G

Solid Mechanics and Aerospace Materials

Chair: Ameer Marzok — Technion
Co-Chair: Pavel Galich — Technion

16:00-16:20	ThL2T7.1

Effect of magnetic field on P- and S-waves in magneto-active polymers...183

Ankush Yadav — Technion
Pavel Galich — Technion

16:20-16:40	ThL2T7.2

An investigation into semi-stabilized unsymmetrical thin-walled structure...184

Steve Katzeff — Israel Aerospace Industries

16:40-17:00	ThL2T7.3

Progressive damage and failure analysis
for structural continuous fiber composites...186

Daniel Vilyatser — Ansys
Tim Artz — Ansys
Eric Stamper — Ansys

17:00-17:20	ThL2T7.4

Multifunction automated unit cell, review...187

Shay Shoam — Israel Aerospace Industries
Uri Ben-Simon — Israel Aerospace Industries
Yuval Freed — Israel Aerospace Industries
Zvi Karuchero — IAI
Alexander Lukatsky — Israel Aerospace Industries
Adam Sawday — IAI
Eduardo Eigenberg — Israel Aerospace Industries
Hilla Elimelech — DDR&D

17:20-17:40	ThL2T7.5

Optimal design of thin-walled beams with buckling considerations...188

Ameer Marzok — Technion

IACAS

Session Halls at the

Faculty of Aerospace Engineering

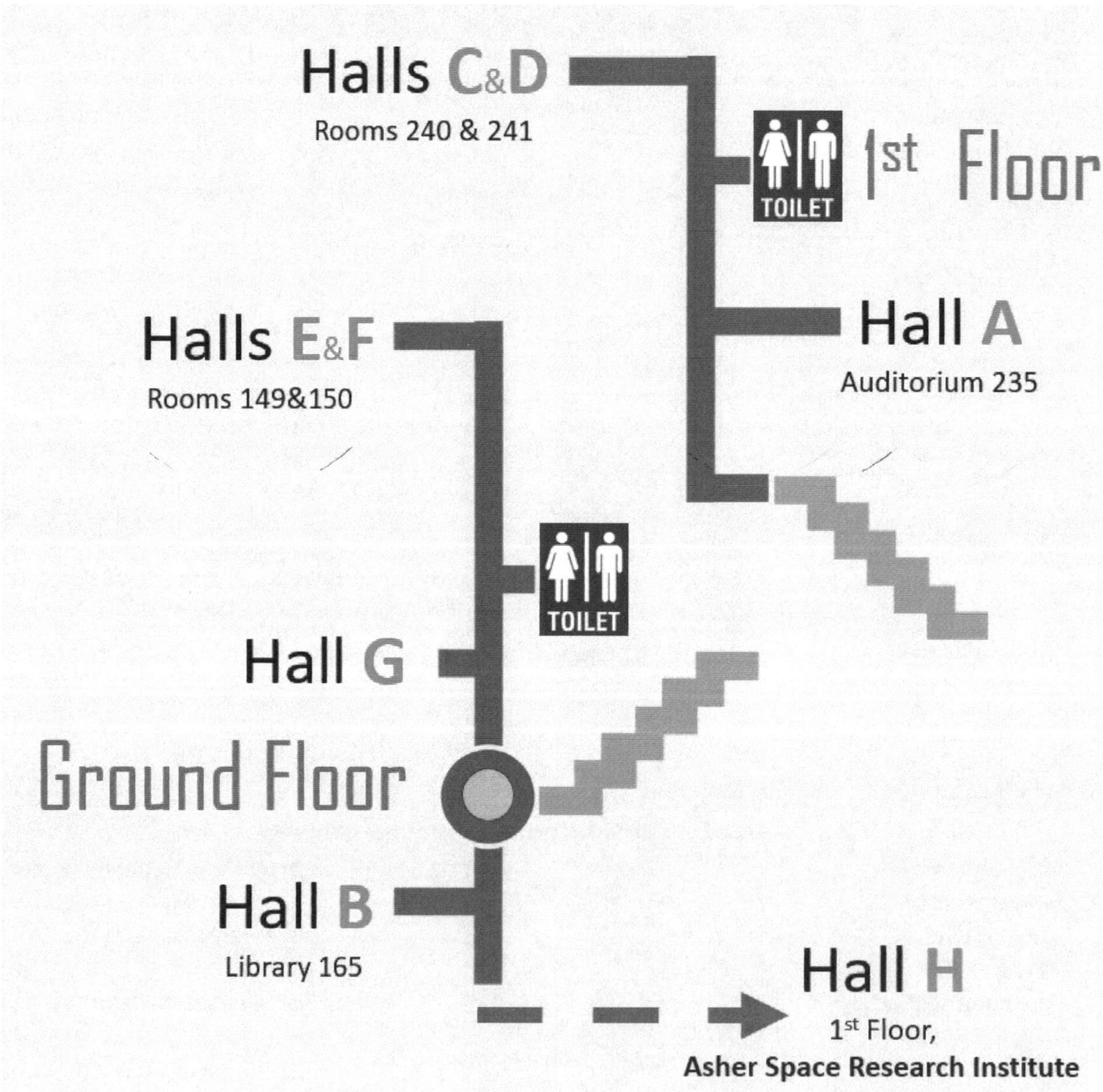

Optimal Pure-Pursuit Guidance Law with Unknown Terminal-Time

Ilan Rusnak, RAFAEL, P.O.Box 2250, Haifa, Israel.

November 28, 2024

1 Extended abstract

The Pure-Pursuit guidance is a well-known guidance law [1, 2]. The guidance principle is based on the desire to aim the velocity of the missile toward the target. It is mechanized by commanding the missile acceleration to be proportional to the angle between the velocity vector and the Line-of-Sight (LOS) to the target. Recently, the optimal control theory has been used to derive optimal Pure-Pursuit guidance law [3]. This guidance law needs the time-to-go. The time-to-go is derivable from the initial range, the missile's velocity profile and an estimate on the target's velocity. Thus, an inherent uncertainty on the value of the terminal-time/time-to-go is created.

This paper tackles the problem of Pure-Pursuit guidance when the time-to-go is a random variable .

The problem of unknown terminal time for quadratic criterion had been solved by [4]. This general solution was applied in [5, 6, 7]. In [8] a guidance law with reduced sensitivity to time-to-go is proposed. Reference [9] presents a stochastic approach to optimal guidance with significant uncertainty in time until intercept.

The general solution [4] is applied to derive optimal Pure-Pursuit guidance laws when the terminal-time is uncertain.

The following is the problem of optimal terminal guidance with uncertain terminal-time [4]. The nth order discrete linear time-varying system is considered

$$x(i+1) = A(i)x(i) + B(i)u(i), x(0) = x_0. \tag{1.1}$$

where $x(i) \in R^n$ is the state, $u(i) \in R^1$ is the input, $A(i) \in R^{nxn}, B(i) \in R^n$.

Optimal guidance laws are derived by posing the following linear quadratic regulator problem on a finite time: given the final time, N, find the function $\{u(i), 0 \leq i \leq N\}$, such that the cost function

Manuscript 14 submitted to 64th Israel Annual Conference on Aerospace Sciences (IACAS). Received December 20, 2024.

$$C_N = \frac{1}{2}\left\{ x(N)^T G_f x(N) + \sum_{i=0}^{N-1} u(i)^T R u(i) \right\} \tag{1.2}$$

is minimized with respect to the input $u(i)$ subject to (1.1) and N (the terminal-time) is known.

The solution of the above problem is well known. Here we consider the case when the final time, N, is a random variable with known distribution $p_N(n)$. The problem being stated here it to find the function $\{u(i), i = 0, 1, 2, ...,\infty\}$ such that the criterion

$$C = E\left[C_N\right] = \frac{1}{2}\sum_{n=0}^{\infty}\left[x(n)^T G_f x(n) + \sum_{i=0}^{n-1} u(i)^T R u(i) \right] p_N(n). \tag{1.3}$$

is minimized. Then following [5] change of summation leads to

$$C = \frac{1}{2}\left\{ \sum_{i=0}^{\infty} \left[x(i)^T \Psi(i) x(i) + u(i)^T \Theta(i) u(i)\right] \right\}. \tag{1.4}$$

Thus the stochastic criterion has been converted into a standard Linear Quadratic Regulator problem with time dependent weights. The solution is

$$\begin{aligned} u(i) &= K(i)x(i), \\ M(i) &= P(i+1)\left[I + B(i)R(i)^{-1}B(i)^T P(i)\right]^{-1}, \\ K(i) &= R(i)^{-1}B(i)^T M(i)A(i), \\ P(i) &= Q(i) + A(i)^T M(i)A(i), \\ P(\infty) &= 0. \end{aligned} \tag{1.5}$$

This is a time-varying Riccati Equation.

The kinematics of a Pure-Pursuit scenario is presented in Figure 1.1.

The linearized equation of motion in the presence of a non-stationary target is

$$\dot{\sigma}_M = \dot{\gamma}_M - \dot{\lambda} = \frac{V_M}{r}\sigma_M - \frac{V_T}{r}\sigma_T + \frac{a_M}{V_M} \tag{1.6}$$

The Velocity Pursuit missile guidance is considered. The objective is to guide the missile toward the target. The miss distance is driven to zero by minimizing the angle between the LOS and the missile's velocity vector, σ_M. This is the objective of the classical Pure-Pursuit (PP) guidance law [1, 10].

$$C = \min_{a_M} \frac{1}{2} \sum_{i=0}^{N_{fmax}} \left\{\Psi(i)\sigma_M^2(i) + \Theta(i)a_M^2(i)\right\} \tag{1.7}$$

Manuscript 14 submitted to 64th Israel Annual Conference on Aerospace Sciences (IACAS). Received December 20, 2024.

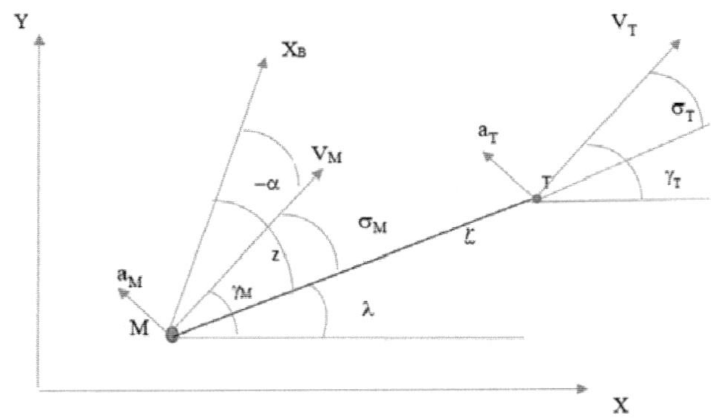

Figure 1.1: The kinematics of a Pure-Pursuit scenario.

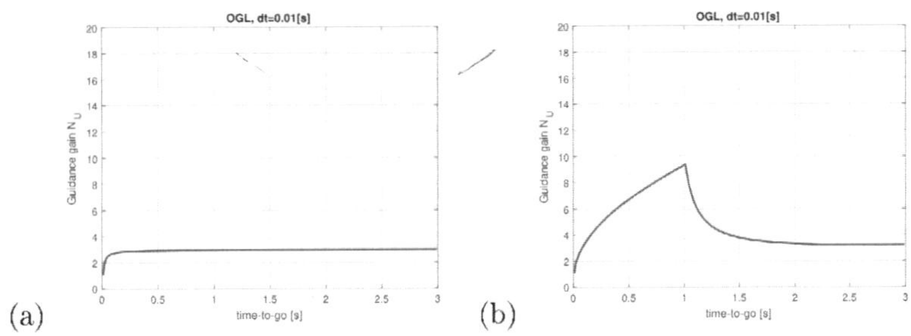

Figure 1.2: The guidance gain .

where N_{fmax} is the expected maximal time of flight.

The optimal PP guidance law for a stationary target with uncertainty of time-to-go is

$$\frac{a_{Mc}(t)}{V_M} = -\frac{N_U(T_{fmax} - t)}{T_{fmax} - t}\sigma_M(t) \tag{1.8}$$

where T_{fmax} is the expected maximal time of flight. Figure (1.2) presents the guidance gain when there is no time-to-go uncertainty(a), $p_N(T_{fmax}) = 1$, and for uniform uncertainty(b) in the last second,

References

[1] N. Shneydor, *Missile guidance and Pursuit-Kinematics, Dynamics and Control.* Horwood Publishing Ltd., 1988.

[2] N.A.Shneydor, *The First Century of Weapon Guidance.* 2nd ed., private publication, 2009.

[3] I. Rusnak, "Optimal pure pursuit missile guidance," *Aerotec. Missili Spaz.*, vol. 103, pp. 51–60, 2024. doi:10.1007/s42496-023-00185-2.

[4] R. Sivan, "The optimal control of linear systems with unknown final time," *IEEE Transactions on Automatic Control*, vol. 11, no. 4, pp. 529–530, 1966. doi:10.1109/TAC.1966.1098388.

[5] I. Rusnak, "Optimal guidance laws with uncertain time-of-flight," *Journal of Guidance, Control and Dynamics*, vol. 33, no. 6, pp. 721–725, 2000. doi: 10.1109/7.845272.

[6] I. Rusnak, "Optimal guidance laws with uncertain time-of-flight against maneuvering target and noisy measurements," *AIAA Guidance, Navigation, and Control Conference and Exhibition*, vol. AIAA 2003-5576, 2003. doi.org/10.2514/6.2003-5576.

[7] I. Rusnak, G. Hexner, and H. Weiss, "Differential games with uncertain terminal time," *IFAC Proceedings Volumes*, vol. 38, no. 1, pp. 182–187, 2005. 16th IFAC World Congress.

[8] J. Ben-Asher and I. Yaesh, *Advances in Missile Guidance Theory.* Vol. 180, American Institute of Aeronautics and Astronautics, Inc., 1998.

[9] G. Hexner and H. Weiss, "Stochastic approach to optimal guidance with uncertain intercept time," *IEEE Transactions on Aerospace and Electronic Systems*, vol. 46, no. 4, pp. 1804–1820, 2010. doi: 10.1109/TAES.2010.5595596.

[10] G. Chatterji and M.Pachter, "Modified velocity pursuit guidance with cross-wind correction against surface targets," *Navigation and Control Conference*, 1991. AIAA-91-2781-CP, doi.org/10.2514/6.1991-2781.

Symmetry-Constrained Formation Maneuvering

Zamir Martinez[1], Daniel Zelazo[2]
[1]Technion Israel Institute of Technology
Email: z.m@campus.technion.ac.il
[2]Technion Israel Institute of Technology
Email: dzelazo@technion.ac.il

Abstract

The demand for advanced coordination schemes in multi-agent systems (MAS) has significantly increased in recent years, driven by applications such as vehicle platooning in autonomous driving [1], drone swarm coordination for surveillance [2], and coordination of satellite constellations for efficient communication relays [3]. While traditional centralized coordination schemes have been widely used for such tasks, they are prone to vulnerability and scalability issues, motivating the exploration of decentralized approaches relying on local sensing and interactions among agents.

This work addresses distance-constrained formation control, a decentralized strategy that enforces desired formations by maintaining specified relative positions and distances between agents. A key challenge in implementing this strategy is achieving a balance between communication requirements and performance constraints. Rigidity theory provides a robust framework for understanding and addressing this trade-off by modeling the decentralized interactions and geometric constraints. Distance constrained rigid frameworks ensure the desired formations by maintaining specific inter-agent distances. Among these, minimal infinitesimal rigid (MIR) frameworks are known to be the minimal architectural requirement that ensures the ensemble converges to the correct shape [4]. In the Euclidean space $\mathbb{R}^2$, MIR translates to having at least $(2n-3)$ communication constraints, where n is the number of agents.

Recent advancements in the theory of symmetry-forced rigidity provided a novel approach that demonstrates the potential of symmetry constrained frameworks as a promising alternative for MIR frameworks [5]. By exploiting specified symmetric constraints found in the formation, this approach notably reduces the required interaction constrains to $(1+1/|\Gamma|)n$, where n is the number of agents and Γ is the underlying symmetry group of the formation. Since symmetry forced formations by definition have point-group symmetries defined with respect to some fixed inertial point, preserving symmetric relationships during motion requires additional considerations.

A key contribution of this work is a modification of the augmented control strategy presented in [5]. This modification introduces a virtual state for each agent to ensure convergence with respect to any arbitrary time-varying centroid, enabling the MAS to achieve and maintain the required spatial pattern during maneuvering. Specifically, consider a network of n agents, described by the gradient-based dynamics:

$$\begin{bmatrix} \dot{p}_0(t) \\ \dot{p}_f(t) \end{bmatrix} = \begin{bmatrix} -\mathcal{O}^T(\mathcal{G}_0, c_0(t))\Big(\mathcal{O}(\mathcal{G}_0, c_0(t))c_0(t) - \mathbf{d}_0^2\Big) \\ 0 \end{bmatrix} - PQP^T \begin{bmatrix} c_0(t) \\ c_f(t) \end{bmatrix}, \tag{1}$$

where $\mathcal{O}$ is the orbit rigidity matrix describing the symmetric infinitesimal rigidity properties of the formation, Q is the symmetry-forcing term, and $\bar{c}(t) = \begin{bmatrix} c_0^T(t) & c_f^T(t) \end{bmatrix}^T = P(p(t) - r(t))$ is the shifted state of the formation according to the virtual state $r(t)$ (see [5]).

Manuscript 56 submitted to 64th Israel Annual Conference on Aerospace Sciences (IACAS). Received January 8, 2025.

By introducing a reference velocity command $v_{ref}(t)$ to the cascade system (1), the system achieves the required formation, rendering the error dynamics exponentially stable while the agents move cohesively according to $v_{ref}(t)$. Fig. 1 illustrates an example of the resulting trajectories when running the implemented control law for $n = 6$ agents. The formation achieves zero steady-state error, as shown in Fig. 2.

Additional results show that by implementing a PI controller in a consensus version of the virtual state, a single leader agent with access to the reference velocity command is sufficient to distributively guide the entire formation while preserving its integrity. Numerical simulations are shown to illustrate the main results.

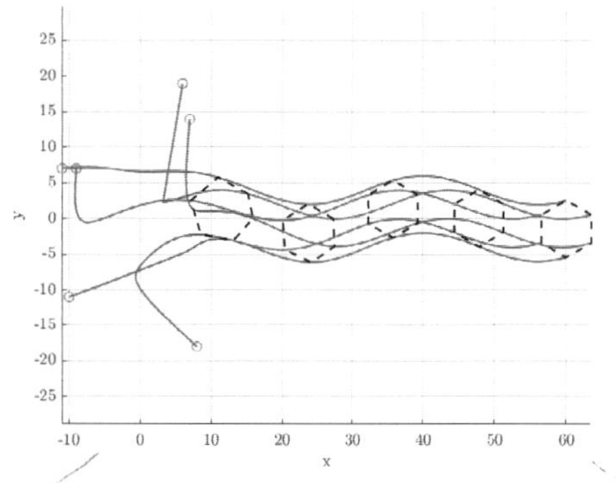

Fig. 1. Maneuvering of a hex-shaped forced symmetric formation according to a reference velocity command $v_{ref}(t) = [1, \pi/4 \cdot \sin(\pi/8 \cdot t)]^T$.

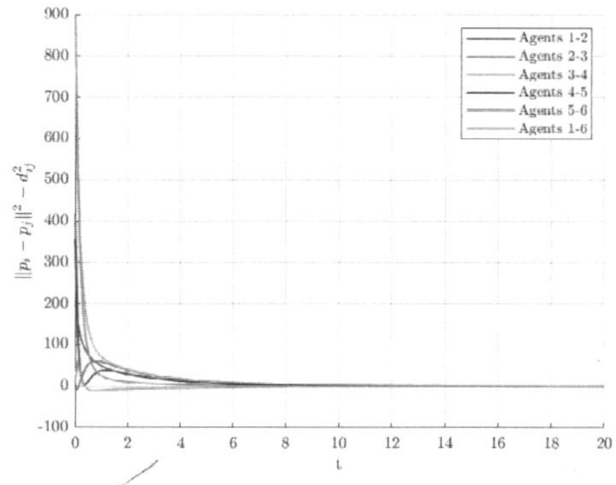

Fig. 2. Error signals $\|p_i - p_j\|^2 - d_{ij}^2$ where d_{ij} is the set of desired inter-agent distances.

References

[1] S.-L. Dai, S. He, H. Lin, and C. Wang, "Platoon formation control with prescribed performance guarantees for usvs," IEEE Transactions on Industrial Electronics, vol. 65, no. 5, pp. 4237–4246, 2018.

[2] S. Martinez and F. Bullo, "Optimal sensor placement and motion coordination for target tracking," Automatica, vol. 42, no. 3, pp. 661–668, 2006.

[3] L. Pedroso and P. Batista, "Distributed decentralized receding horizon control for very large-scale networks with application to satellite mega-constellations," arXiv, 2023.

[4] L. Krick, M. E. Broucke, and B. A. Francis, "Stabilisation of infinitesimally rigid formations of multi-robot networks," International Journal of Control, vol. 82, no. 3, pp. 423–439, 2009.

[5] D. Zelazo, S. ichi Tanigawa, and B. Schulze, "Forced symmetric formation control," arXiv, 2024.

Extended Abstract

Missile Guidance with Doppler Information Only in 3D

Liat Peled-Eitan and Ilan Rusnak
RAFAEL, POBox 2250, Haifa, 3102102, Israel

Target tracking and missile guidance in 3D applicable with EM sensor is presented and analyzed. Only the Doppler frequency measurement is used. The fact that the Doppler frequency is maximal when the missile velocity is aligned with line-of-sight is used. The steepest descent zig-zag algorithm [4] for maximizing the closing velocity is used. This realizes a Pure-Pursuit guidance law [3]. The feasibility of the Doppler-based guidance is shown and demonstrated by simulation in a 3D scenario against a stationary target.

The problem of tracking a target by a moving platform based on Doppler frequency measurement only is presented is analyzed in [1,2]. A method for tracking and guidance in 2D is documented in [5]. The idea is as follows: Assume a moving missile with an appropriate EM sensor that measures the Doppler frequency of a stationary target. It is well known that the Doppler frequency is at maximum when the missile moves along the line-of-sight (LOS). Thus, if the missile velocity is kept aligned with the LOS, the target is tracked, the missile is guided toward the target, and the scenario ends in an encounter, i.e., intercept. Thus, only the Doppler frequency and angular measurement are not required. The scenario considered in this paper is presented in Figure 1.

The equations of motion in the plane defined by the missile velocity, V_M, and the LOS vector, λ, are

$$\begin{aligned} \sigma_M &= \gamma_M - \lambda \\ \dot{r}_{TM} &= -V_M \cos(\sigma_M) \\ r_{TMAZ}\dot{\lambda} &= -V_M \sin(\sigma_M) \\ \dot{\gamma}_M &= \frac{a_M}{V_M} \end{aligned} \tag{1}$$

where a_M is the missile's acceleration in the missile velocity-LOS plane.

The missile considered in this paper is roll-stabilized and has two independent lateral guidance channels. The aspect angles and accelerations in the missile coordinate frame are σ_{MAz}, a_{MAZ} in the $x - z$ plane and σ_{MEl}, a_{MEl} in the x-y plane, respectively, as depicted in Figure 2.

Manuscript 20 submitted to 64th Israel Annual Conference on Aerospace Sciences (IACAS). Received December 22, 2024.

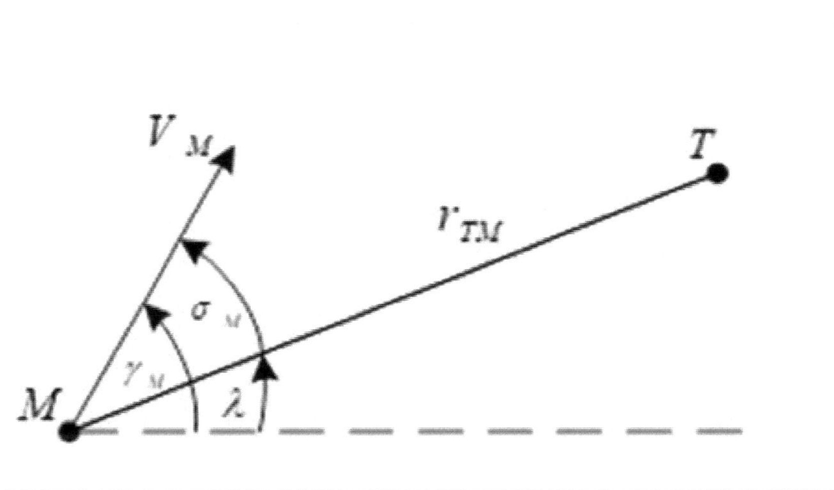

Figure 1: The kinematics for a stationary target in the plane defined by the missile velocity, V_M, and the LOS vector.

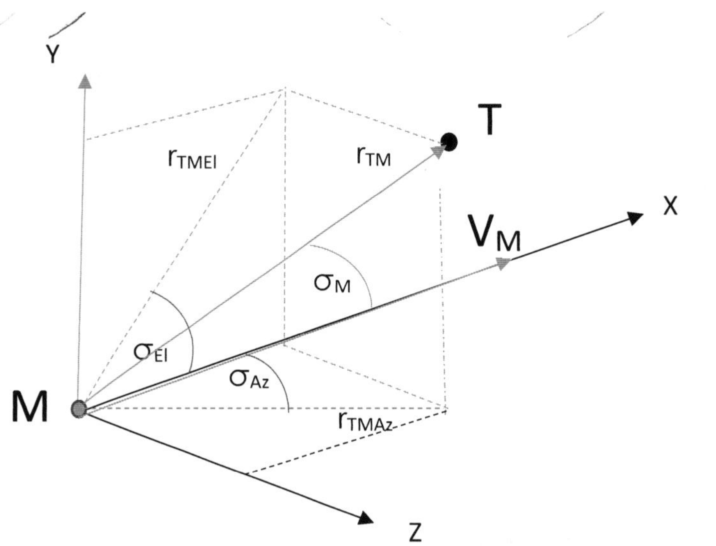

Figure 2: The missile coordinates frame

The equations of motion in the two perpendicular axes of the missile, El and $Az,$ can be approximated (This is a crude approximation. It is presented for simplicity.) as

$$\dot{\sigma}_{MEl} = \frac{V_M}{r_{TMEl}} \sin(\sigma_{MEl}) + \frac{a_{MEl}}{V_M} \tag{3}$$

$$\dot{\sigma}_{MAz} = \frac{V_M}{r_{TMAz}} \sin(\sigma_{MAz}) + \frac{a_{MAz}}{V_M}. \tag{4}$$

Further, we have

$$a_M = a_{MEl}\sin\theta + a_{MAz}\cos\theta\,, \qquad \tan\theta = \frac{a_{MEl}}{a_{MAz}}. \tag{5}$$

$$a_M = \sqrt{a_{aMEl}^2 + a_{MAz}^2} \tag{6}$$

To guide the missile toward the target we drive the angle between the LOS and the missile's velocity, σ_M, to zero.

The moving missile's sensor measures the Doppler that is for passive sensing as depicted in Figure 2 and in Figure 3 that shows equal closing velocity lines

$$f_d = -\frac{\dot{r_{TM}}}{\lambda} = -\frac{\dot{\rho}}{\lambda} = \frac{V_M}{\lambda}\cos(\sigma_M) \tag{8}$$

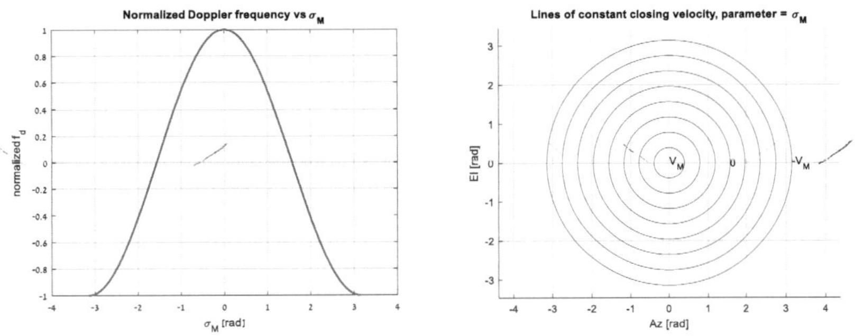

Figure 3: The normalized Doppler frequency, f_d, versus σ_M.

The Gradient Descent/Ascent zig-zag optimization algorithm [4] is applied. The optimal Pure Pursuit is implemented [6].

Figure 4a presents the missile trajectory and accelerations in 3D. Figures 4c and d show the trajectory projections the on x-y and x-z planes, respectively. The range rate, aspect angle and missile's azimuth and elevation accelerations are depicted in Figure 4b. Figure 4e gives the aspect angle in the missile's coordinates, i.e. the y-z plane.

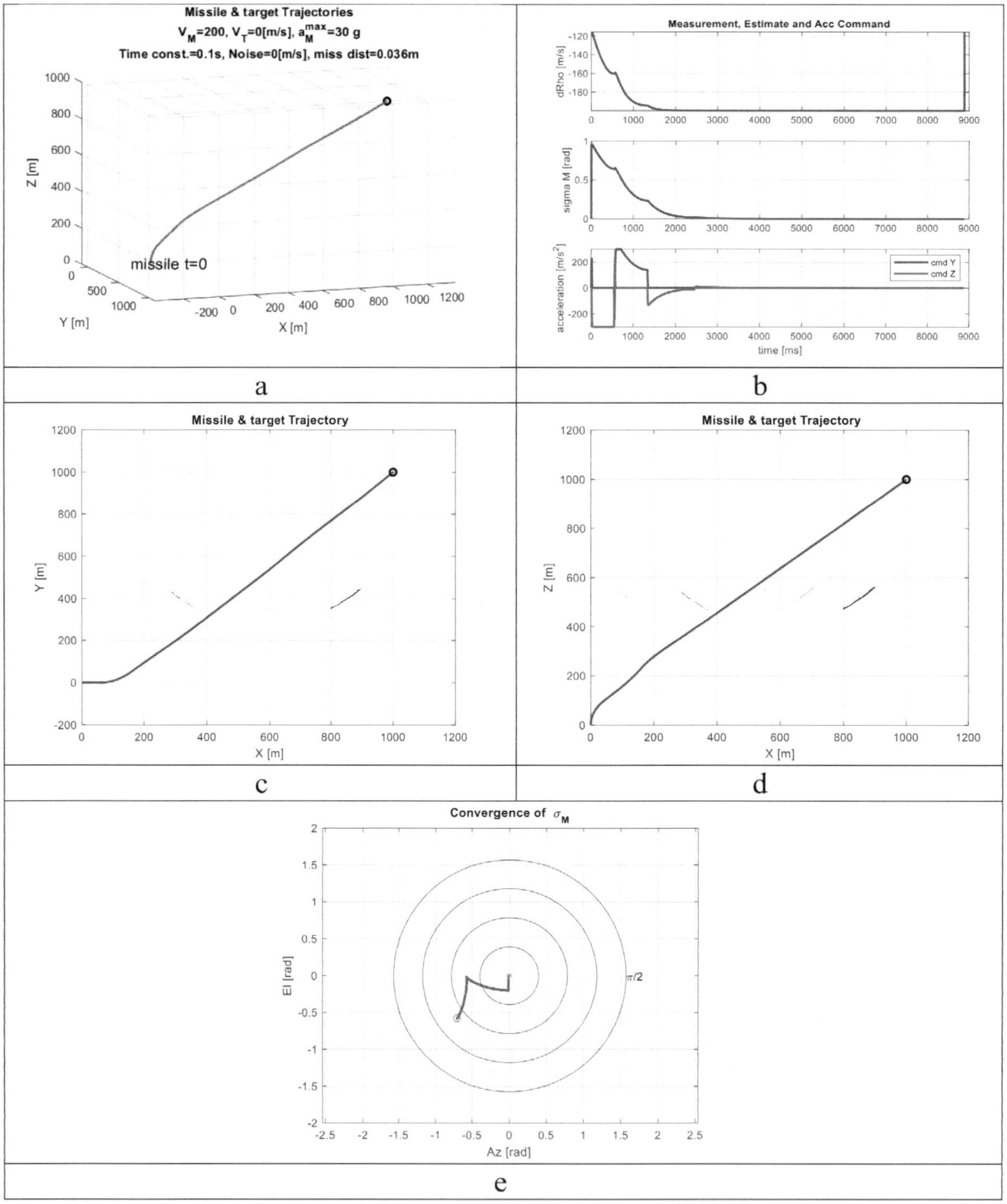

Figure 4: The missile trajectory, accelerations and the aspect angle in 3D.

References

1] Qin Yu-Liang, Wang Jian-Tao, and Deng Bin: Range-Doppler Measurements Based Target Locating for Active Radar Seekers, 2009 2nd Asian-Pacific Conference on Synthetic Aperture Radar, Xian, Shanxi, China, 26-30 Oct. 2009. DOI: 10.1109/APSAR.2009.5374198

2] Liat Peled-Eitan and Ilan Rusnak: Tracking a stationary target by moving platform with Doppler information only, the 60th Israel Annual Conference on Aerospace Science, March 4-5, 2020, Israel.

3] Shneydor, N.A.: Missile Guidance and Pursuit: Kinematics, Dynamics and Control, Horwood Publishing Limited, 1998.

4] Shashi Kant Mishra and Bhagwat Ram: Introduction to Unconstrained Optimization with R, Springer. doi.org/10.1007/978-981-15-0894-3

5] Liat Peled-Eitan and Ilan Rusnak: Missile Guidance on Stationary Target with Doppler Information Only, 10th International Conference On Control, Decision And Information Technologies, CoDIT 2024, 01-04 July, 2024, Valetta, Malta.

6] Ilan Rusnak: Optimal Pure Pursuit Missile Guidance, Aerotecnica Missili & Spazio, 20 October 2023. DOI: 10.1007/s42496-023-00185-2

Manuscript 20 submitted to 64th Israel Annual Conference on Aerospace Sciences (IACAS). Received December 22, 2024.

A New Atmospheric Disturbances Estimation Method for Rotorcraft

The development of civilian unmanned aircraft systems (UAS) has introduced problems related to flight safety and economic performance. Atmospheric wind significantly affects autonomous rotorcrafts' navigation and flight safety. Thus, estimating the wind conditions in flight is critical. Such estimates can be used to modify guidance laws, downlinked to the ground control station, and shared with other aircraft.

The classic method for fixed-wing aircraft consists of using the wind triangle. The wind can be estimated if the true airspeed (TAS) is known from the air data system and the ground speed vector is known from the navigation system. However, most small rotorcrafts are not equipped with angle-of-attack and side-slip sensors. The assumption that the measured indicated airspeed (IAS) is approximately equal to TAS may be valid for fixed-wing aircraft. However, it is unacceptable for a single-rotor helicopter that can fly with significant slip angles.

This study continues the authors' research with the addition of new practical outcomes. The research proposes a unique model-based algorithm for estimating atmospheric disturbances using the inertial navigation system and Pitot measurements. This approach does not involve special maneuvers and does not assume steady flight like other existing methods. The BE-50 of Steadicopter Ltd, Israel, is chosen as the helicopter flight test platform.

A linear identified model is proposed as a basic model of helicopter dynamics, which is used for flight control design and is available for all UAS projects. Linear models were identified and verified for several anchor points within the flight envelope. Two linear helicopter models for hover and forward flight were selected and modified by adding trim data and nonlinear kinematic equations. A nonlinear Unscented Kalman filter was designed to estimate wind components in the local inertial frame.

This algorithm has been tested on a simulation, which includes the wind and atmospheric turbulence simulated by the Dryden model. The result was validated on the BE-50 helicopter flight tests. Two types of flight tests were performed. The first test was a flight with constant airspeed. The second test was performed for hover conditions. In this last experiment, wind characteristics were also measured using an ultrasonic wind sensor. The results were compared with the estimate obtained using the Unscented Kalman filter.

Manuscript 78 submitted to 64th Israel Annual Conference on Aerospace Sciences (IACAS). Received January 12, 2025.

In the future, the current work will be extended and modified by continuing flight testing and improving the gust matrix. We will also research the connection between identifying actual atmospheric disturbances and identifying the control equivalent turbulent input (CETI).

Error Analysis of a Dual Quaternion Batch Estimator

Caitong Peng★ *and Daniel Choukroun*†
Ben-Gurion University of the Negev, POB 653
Beer-Sheva, 84105, Israel
caitongpeng@gmail.com · danielch@bgu.ac.il
†Corresponding author

Abstract

This study presents an error analysis for a single-frame dual quaternion batch estimator that utilizes point and unit vector measurements. This estimator addresses a constrained least-squares problem by minimizing a cost function that aggregates orientation and position errors, analogous to the q-method for rotation quaternions. An error analysis of this dual quaternion method is developed based on the eigenvalue principle for the first time, deriving the closed-form expressions for the three-dimensional attitude and translation error vectors, along with their covariance matrices. Extensive Monte Carlo simulations validate the precision of these expressions. Moreover, we explore the algorithm's sensitivity to variations in batch sizes and cost weights, revealing the balance between computational efficiency and estimation accuracy.

1. Introduction

Dual quaternions combine Hamilton's quaternion theory[1] and Clifford's dual number theory[2] to represent the pose, i.e., the rotation and the translation in rigid-body kinematics. Compared to the traditional transformation matrices, dual quaternions eliminate the need for complex matrix multiplications, significantly reducing computational load and storage demands. Their eight parameters are an integrated representation of six degrees of freedom (6-DOF), with some redundancy. This integration facilitates the direct computation of statistical cross-covariances between rotational and translational movements and enhances the precision and effectiveness of spatial analyses. Detailed studies of dual quaternions in three-dimensional kinematics were provided by[3]. A comprehensive analysis of dual numbers and their applications in the kinematic analysis of spatial linkages was conducted in[4], which has since become widely adopted. Subsequently, Bottema and Roth[5] studied theoretical kinematics using dual numbers. Dooley and McCarthy[6] utilized dual quaternions to generalize coordinates to characterize the dynamics of six degrees of freedom (6-DOF) in a single rigid body. An extensive exploration of dual number theory, its application to dynamic system modeling, and handling functions involving dual variables were presented in[7]. In addition, dual quaternions have also been widely applied in fields such as kinematic analysis and computer graphics, as demonstrated by the work of Angeles[8], and Konstantinos[9].

The application of dual quaternions to depict pose in 6-DOF systems has gained significant attention in the last two decades. In aerospace applications, particularly for spacecraft performing maneuvers, dual quaternions can be used to model, estimate, and control spatial orientation and position. Moreover, dual quaternions excel at representing multiple satellites' relative positions and attitudes in collaborative operations or formation flying. In the broad field of single-frame pose estimation, least-squares optimization techniques are often used to find the best estimate of the transformation matrix. One approach prioritizes optimizing the orientation (rotation matrix) before determining the translation[10],[11]. That approach may lead to compounded errors due to error propagation from inaccuracies in the rotation matrix and the corresponding data points. Another method simultaneously tackles the tasks of orientation and position estimation at the expense of efficiency[12]. Both strategies reflect some separation in handling the transformation matrix's rotational and translational elements. In[13], an efficient algorithm was introduced that utilizes dual quaternions to simultaneously determine an object's orientation and position. That method minimizes a cost function that captures the sum of orientation and position errors. It outperforms previous methods in terms of accuracy and speed. Initially devised as a naive least-squares optimization under constraints, that method has not been thoroughly analyzed regarding its deterministic and random errors. In particular, no expressions for the covariance matrix of the estimation error have been developed up to now. The contribution of this research is a thorough error analysis of the single-frame dual quaternion estimator from points and vector observations presented in[13]. Based on the eigenvalue principle, the analysis yields expressions for

Manuscript 58 submitted to 64th Israel Annual Conference on
Aerospace Sciences (IACAS). Received January 9, 2025.

the three-dimensional attitude and translation error vectors and their covariance matrices. The validity of these formulas is shown through extensive Monte Carlo simulations. The dual-quaternion estimator's sensitivity to the size and weighting of the measurement set is numerically evaluated, providing critical insights into its practical applicability and robustness.

2. Dual Number Quaternions

William Kingdom Clifford introduced the concept of dual quaternions in 1873[14], which is an extension that integrates Hamilton's quaternion theory[1] with dual numbers, facilitating the unified representation of rotations and translations in rigid-body kinematics. A dual quaternion $\mathcal{q}$ is defined as:

$$\mathcal{q} = \boldsymbol{q} + \boldsymbol{q}_d \tag{1}$$

where $\boldsymbol{q}$ and $\boldsymbol{q}_d$ are the real part and dual part, both being quaternions. For two dual quaternions $\mathcal{q}_1 = \boldsymbol{q}_1 + \epsilon \boldsymbol{q}_{d1}$ and $\mathcal{q}_2 = \boldsymbol{q}_2 + \epsilon \boldsymbol{q}_{d2}$ are dual quaternions, the dual quaternion addition and multiplication are defined as follows:

$$\mathcal{q}_1 + \mathcal{q}_2 = (\boldsymbol{q}_1 + \boldsymbol{q}_2) + \epsilon(\boldsymbol{q}_{d1} + \boldsymbol{q}_{d2}) \tag{2}$$

$$\mathcal{q}_1 \odot \mathcal{q}_2 = [\mathcal{q}_1 \odot]\mathcal{q}_2 = \begin{bmatrix} [\boldsymbol{q}_1\odot] & 0_4 \\ [\boldsymbol{q}_{d1}\odot] & [\boldsymbol{q}_1\odot] \end{bmatrix} \begin{bmatrix} \boldsymbol{q}_2 \\ \boldsymbol{q}_{d2} \end{bmatrix} = \begin{bmatrix} [\boldsymbol{q}_1\odot]\boldsymbol{q}_2 \\ [\boldsymbol{q}_1\odot]\boldsymbol{q}_{d2} + [\boldsymbol{q}_{d1}\odot]\boldsymbol{q}_2 \end{bmatrix} \tag{3}$$

$$= [\odot \mathcal{q}_2]\mathcal{q}_1 = \begin{bmatrix} [\odot\boldsymbol{q}_2] & 0_4 \\ [\odot\boldsymbol{q}_{d2}] & [\odot\boldsymbol{q}_2] \end{bmatrix} \begin{bmatrix} \boldsymbol{q}_1 \\ \boldsymbol{q}_{d1} \end{bmatrix} = \begin{bmatrix} [\odot\boldsymbol{q}_2]\boldsymbol{q}_1 \\ [\odot\boldsymbol{q}_{d2}]\boldsymbol{q}_1 + [\odot\boldsymbol{q}_2]\boldsymbol{q}_{d1} \end{bmatrix} \tag{4}$$

Here, $\odot$ denotes both the dual quaternion multiplication and the quaternion multiplication, with $[\boldsymbol{q}\odot]$ and $[\odot\boldsymbol{q}]$ are the quaternion product matrices:

$$[\boldsymbol{q}\odot] = \begin{bmatrix} [\mathbf{e_r}\times] + \mathrm{q_r}\mathrm{I}_3 & \mathbf{e_r} \\ -\mathbf{e_r^T} & \mathrm{q_r} \end{bmatrix} \tag{5}$$

$$[\odot\boldsymbol{q}] = \begin{bmatrix} -[\mathbf{e_r}\times] + \mathrm{q_r}\mathrm{I}_3 & \mathbf{e_r} \\ -\mathbf{e_r^T} & \mathrm{q_r} \end{bmatrix} \tag{6}$$

One of the conjugates of $\mathcal{q}$ follows from the classical quaternion conjugation:

$$\mathcal{q}^* = \boldsymbol{q}^* + \epsilon\boldsymbol{q}_d^* = [\mathrm{q}_0, -\mathrm{q}_1, -\mathrm{q}_2, -\mathrm{q}_3, \mathrm{q}_{d0}, -\mathrm{q}_{d1}, -\mathrm{q}_{d2}, -\mathrm{q}_{d3}] \tag{7}$$

A dual quaternion $\mathcal{q} = \boldsymbol{q} + \epsilon\boldsymbol{q}_d$ is unit if $\mathcal{q} \odot \mathcal{q}^* = 1$. The unit dual quaternion adheres two constraints:

$$\boldsymbol{q}^T\boldsymbol{q} = 1 \tag{8}$$

$$\boldsymbol{q}_d^T\boldsymbol{q} = 0 \tag{9}$$

These two constraints ensure that dual quaternions can unambiguously represent rigid body transformations in three-dimensional space, with the degrees of freedom effectively reduced to six.
Chasles' theorem, formulated in 1830, articulates that the motion of a rigid body in three-dimensional space can be decomposed into a rotation about an axis and a translation along the same axis[15]. This transformation is depicted in Fig. 1, where the reference coordinate system $\mathcal{R}$ rotates around an axis defined by the unit vector **n** and point **p**, through an angle θ, followed by a translation by distance d along **n**, resulting in the body coordinate system $\mathcal{B}$.

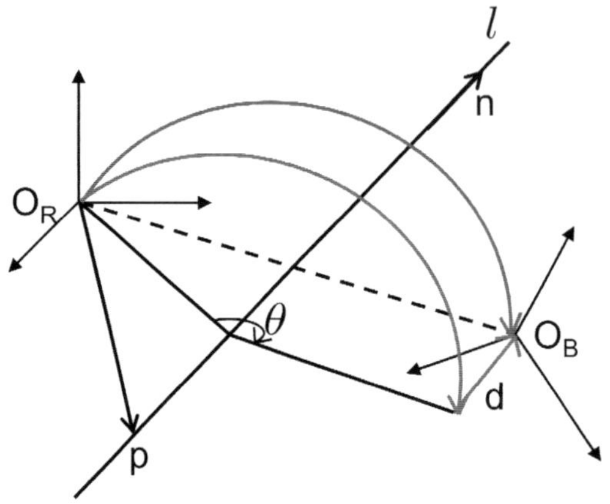

Figure 1: Schematics of the Screw Motion

Dual quaternions provide a mathematical framework that is particularly effective to describe the screw motions. The rotation component of the dual quaternion $\boldsymbol{q}$ and its dual counterpart $\boldsymbol{q_d}$ are expressed as:

$$\boldsymbol{q} = \begin{bmatrix} \sin(\theta/2)\mathbf{n} \\ \cos(\theta/2) \end{bmatrix} \tag{10}$$

$$\boldsymbol{q_d} = \begin{bmatrix} (d/2)\cos(\theta/2)\mathbf{n} + \sin(\theta/2)(\mathbf{p}\times\mathbf{n}) \\ -(d/2)\sin(\theta/2) \end{bmatrix} \tag{11}$$

3. A Single-frame Dual Quaternion Estimator

This section describes a batch estimator of a rigid body pose introduced in[13] and developed as a constrained least-square problem. The estimation problem is formulated as the minimization of a least-squares cost designed using measurement residuals from point and vector observations, under the constraints inherent to unit dual quaternions. Compared to conventional methods, the dual quaternion technique is more efficient in processing sensor inputs, such as object surface points and directional vectors. For ease of analysis, we use quaternion notation to represent these quantities. Let $\mathbf{p}$ and $\boldsymbol{p}$ denote the position vector of a point on the object's surface and the associated quaternion, respectively, as follows:

$$\boldsymbol{p} = \frac{1}{2}\begin{bmatrix} \mathbf{p} \\ 0 \end{bmatrix} \tag{12}$$

Similarly, let $\mathbf{n}$ and $\boldsymbol{n}$ denote a unit vector and its associated quaternion, respectively, as follows:

$$\boldsymbol{n} = \begin{bmatrix} \mathbf{n} \\ 0 \end{bmatrix} \tag{13}$$

Assume that l points and k unit vectors are measured and body-referenced to the frame $\mathcal{B}$. Let $\boldsymbol{p}_{b_i}$ and $\boldsymbol{n}_{b_i}$ denote the corresponding quaternion measurements in $\mathcal{B}$, respectively. Each measurement, point, or unit vector is associated with a similar quantity known in a reference frame, $\mathcal{R}$. Let $\boldsymbol{p}_{r_i}$ and $\boldsymbol{n}_{r_i}$ denote these reference frame quaternions. The frame transformation from $\mathcal{R}$ to $\mathcal{B}$ for the points and the unit vectors are expressed as follows:

$$\boldsymbol{p}_{b_i} = W(\boldsymbol{q})^T\boldsymbol{q_d} + W(\boldsymbol{q})^T Q(\boldsymbol{q})\boldsymbol{p}_{r_i} \tag{14}$$

$$\boldsymbol{n}_{b_i} = W(\boldsymbol{q})^T Q(\boldsymbol{q})\boldsymbol{n}_{r_i} \tag{15}$$

where $\boldsymbol{q}$ denotes the rotation quaternion and $\boldsymbol{q_d}$ denotes the dual part of the dual quaternion from $\mathcal{R}$ to $\mathcal{B}$. The quaternion $\boldsymbol{q_d}$ is readily calculated using $\boldsymbol{q}$ and the translation vector from the origin of $\mathcal{B}$ to the origin of $\mathcal{R}$. Let $\boldsymbol{q}$ consist of a vector part $\mathbf{e} = [e_1, e_2, e_3]^T$ and a scalar part q, then the matrices $W(\boldsymbol{q})$ and $Q(\boldsymbol{q})$ are defined as follows:

$$Q(\boldsymbol{q}) = \begin{bmatrix} qI + K(\mathbf{e}) & \mathbf{e} \\ -\mathbf{e}^T & q \end{bmatrix} \tag{16}$$

$$W(\boldsymbol{q}) = \begin{bmatrix} qI - K(\mathbf{e}) & \mathbf{e} \\ -\mathbf{e}^T & q \end{bmatrix} \tag{17}$$

Manuscript 58 submitted to 64th Israel Annual Conference on Aerospace Sciences (IACAS). Received January 9, 2025.

and $K(\mathbf{e})$ is the skew-symmetric matrix defined as follows:

$$K(\mathbf{e}) = \begin{bmatrix} 0 & -\mathrm{e}_3 & \mathrm{e}_2 \\ \mathrm{e}_3 & 0 & -\mathrm{e}_1 \\ -\mathrm{e}_2 & \mathrm{e}_1 & 0 \end{bmatrix} \tag{18}$$

A constrained least-squares estimation problem is formulated by considering the following cost:

$$E = \sum_{i=1}^{k} \alpha_i \|\boldsymbol{n}_{r_i} - \boldsymbol{n}_{b_i}\|^2 + \sum_{i=1}^{l} \beta_i \|\boldsymbol{p}_{r_i} - \boldsymbol{p}_{b_i}\|^2 \tag{19}$$

where α_i and β_i are positive weights, and minimizing it is subject to two constraints:

$$\boldsymbol{q}^T\boldsymbol{q} = 1$$
$$\boldsymbol{q}^T\boldsymbol{q_d} = 0$$

Substituting Eqs. (14),(15) into Eq. (19) yields:

$$E = \boldsymbol{q}^T C_1 \boldsymbol{q} + \boldsymbol{q_d}^T C_2 \boldsymbol{q_d} + \boldsymbol{q_d}^T C_3 \boldsymbol{q} + C_4 \tag{20}$$

where

$$C_1 = -2\sum_{i=1}^{k} \alpha_i Q(\boldsymbol{n}_{b_i})^T W(\boldsymbol{n}_{r_i}) - 2\sum_{i=1}^{l} \beta_i Q(\boldsymbol{p}_{b_i}) W(\boldsymbol{p}_{r_i})$$
$$C_2 = \left(\sum_{i=1}^{l} \beta_i\right) I$$
$$C_3 = 2\sum_{i=1}^{l} \beta_i (W(\boldsymbol{p}_{r_i}) - Q(\boldsymbol{p}_{b_i}))$$
$$C_4 = 2\sum_{i=1}^{k} \alpha_i + \sum_{i=1}^{l} \beta_i \left((\boldsymbol{p}_{r_i}^T \boldsymbol{p}_{r_i} + \boldsymbol{p}_{b_i}^T \boldsymbol{p}_{b_i})\right)$$

The error function is then minimized without constraints:

$$\tilde{E} = \boldsymbol{q}^T C_1 \boldsymbol{q} + \boldsymbol{q_d}^T C_2 \boldsymbol{q_d} + \boldsymbol{q_d}^T C_3 \boldsymbol{q} + C_4 + \lambda_1(\boldsymbol{q}^T\boldsymbol{q} - 1) + \lambda_2(\boldsymbol{q_d}^T \boldsymbol{q}) \tag{21}$$

where λ_1 and λ_2 are Lagrange multipliers. Deriving the partial derivatives yields:

$$\frac{\partial \tilde{E}}{\partial \boldsymbol{q}} = (C_1 + C_1^T)\boldsymbol{q} + C_3^T \boldsymbol{q_d} + 2\lambda_1 \boldsymbol{q} + \lambda_2 \boldsymbol{q_d} = 0 \tag{22}$$
$$\frac{\partial \tilde{E}}{\partial \boldsymbol{q_d}} = (C_2 + C_2^T)\boldsymbol{q_d} + C_3 \boldsymbol{q} + \lambda_2 \boldsymbol{q} = 0 \tag{23}$$

Multiplying Eq. (23) by $\boldsymbol{q}$ from the left and noting that C_3 is skew-symmetric yields

$$\lambda_2 = 0 \tag{24}$$

Then Eq. (23) yields the following relationship between $\boldsymbol{q}$ and the dual part $\boldsymbol{q_d}$:

$$\boldsymbol{q_d} = -(C_2 + C_2^T)^{-1} C_3 \boldsymbol{q} \tag{25}$$

Substituting Eq. (25) into Eq. (22) yields:

$$A\boldsymbol{q} = \lambda_1 \boldsymbol{q} \tag{26}$$

where

$$A = \left(\frac{1}{2} C_3 (C_2 + C_2^T) C_3^T - C_1 - C_1^T\right) \tag{27}$$

Manuscript 58 submitted to 64th Israel Annual Conference on
Aerospace Sciences (IACAS). Received January 9, 2025.

Equation. (27) states that the quaternion $\boldsymbol{q}$ is an eigenvector of the matrix A associated with the eigenvalue λ_1. Multiplying Eqs. (22) and (23) by $\boldsymbol{q}^T$ and $\boldsymbol{q_d}^T$, respectively, yields:

$$\boldsymbol{q}^T C_1 \boldsymbol{q} = \frac{1}{2}\boldsymbol{q}^T (C_1 + C_1{}^T)\boldsymbol{q} = -\frac{1}{2}\boldsymbol{q_d}^T C_3 \boldsymbol{q} - \lambda_1 \tag{28}$$

$$\boldsymbol{q_d}^T C_2 \boldsymbol{q_d} = \frac{1}{2}\boldsymbol{q_d}^T (C_2 + C_2{}^T)\boldsymbol{q_d} = -\frac{1}{2}\boldsymbol{q_d}^T C_3 \boldsymbol{q} \tag{29}$$

Substituting these into Eq. (20) gives:

$$\tilde{E} = C_4 - \lambda_1 \tag{30}$$

Thus, the cost is minimized by choosing $\boldsymbol{q}$ as the eigenvector of A associated with the largest positive eigenvalue.

Summary of the dual quaternion estimator

Given batches of point quaternion observations, $\{\boldsymbol{p}_{b_i}, \boldsymbol{p}_{r_i}\}_{i=1}^{l}$, and unit vector quaternion observations, $\{\boldsymbol{n}_{b_i}, \boldsymbol{n}_{r_i}\}_{i=1}^{k}$, compute the following four-dimensional matrices:

$$C_1 = -2\sum_{i=1}^{k} \alpha_i Q(\boldsymbol{n}_{b_i})^T W(\boldsymbol{n}_{r_i}) - 2\sum_{i=1}^{l} \beta_i Q(\boldsymbol{p}_{b_i}) W(\boldsymbol{p}_{r_i})$$

$$C_2 = \left(\sum_{i=1}^{l} \beta_i\right) I$$

$$C_3 = 2\sum_{i=1}^{l} \beta_i (W(\boldsymbol{p}_{r_i}) - Q(\boldsymbol{p}_{b_i}))$$

and the matrix A as follows:

$$A = \frac{1}{2} C_3 (C_2 + C_2{}^T) C_3{}^T - (C_1 + C_1{}^T) \tag{31}$$

Then, extract the optimal rotation quaternion estimate $\boldsymbol{q}$ as the eigenvector of A associated with the maximal eigenvalue, and compute the dual part $\boldsymbol{q_d}$ as follows:

$$\boldsymbol{q_d} = -(C_2 + C_2{}^T)^{-1} C_3 \boldsymbol{q} \tag{32}$$

This algorithm is very similar to the q-method; it extracts the rotation quaternion as the eigenvector of a four-dimensional real symmetric matrix that is derived from the data. Indeed, in the case where only unit vector measurements are acquired, the problem formulation is formally identical to the Wahba problem.

4. Error Analysis

This section conducts an error analysis of the algorithm that was previously outlined. We first derive expressions for the rotation error vector $\boldsymbol{\delta\theta}$ and the translation error vector $\mathbf{\Delta t}$. Then, we perform a random error analysis to ascertain the covariance matrix associated with these error vectors.

4.1 Pose Error Vector

Measurement noise significantly affects data accuracy in both the reference frame and the spacecraft's body frame, influencing position and orientation calculations. We parameterize the attitude error using a rotation vector $\boldsymbol{\delta\theta}$ and the position error using a translation vector $\mathbf{\Delta t}$. These vectors represent the discrepancies in rotation and translation between the estimated body frame $\hat{\mathcal{B}}$ and the true body frame $\mathcal{B}$.

$$\hat{\mathcal{q}} = \mathcal{q}_{RB} \odot \mathcal{q}_{B\hat{B}} = (\boldsymbol{q}' + \epsilon \boldsymbol{q_d}') \odot (\delta\boldsymbol{q} + \epsilon\delta\boldsymbol{q_d}) = (\boldsymbol{q} \odot \delta\boldsymbol{q}) + \epsilon(\boldsymbol{q_d} \odot \delta\boldsymbol{q} + \boldsymbol{q} \odot \delta\boldsymbol{q_d}) \tag{33}$$

where $\odot$ means dual quaternion multiplication and quaternion multiplication. The error analysis utilizes first-order approximations in measurement errors. Eq. 33 can be approximated using Taylor series expansions as follows:

$$\hat{\mathcal{q}} = (\boldsymbol{q}' + \boldsymbol{q}' \odot \begin{bmatrix} \frac{1}{2}\boldsymbol{\delta\theta} \\ 0 \end{bmatrix}) + \epsilon(\boldsymbol{q_d}' + \boldsymbol{q_d}' \odot \begin{bmatrix} \frac{1}{2}\boldsymbol{\delta\theta} \\ 0 \end{bmatrix} + \boldsymbol{q}' \odot \begin{bmatrix} \frac{1}{2}\mathbf{\Delta t} \\ 0 \end{bmatrix} \odot \boldsymbol{q}') \tag{34}$$

where superscript ^{t} denotes the true value of the underlying variable. We are only considering the multiplicative error of the dual quaternion and not the additive error, and given that the additive noise of unit vectors and points is very small, These two points allow us to assume that the reference frame $\mathcal{R}$ and body frame $\mathcal{B}$ coincide. With this assumption, the true values of the dual quaternion can be simplified and Eq. 34 yields:

$$\hat{q} = (\begin{bmatrix}\mathbf{0}_3\\1\end{bmatrix} + \frac{1}{2}\begin{bmatrix}\delta\boldsymbol{\theta}\\0\end{bmatrix}) + \epsilon(\frac{1}{2}\begin{bmatrix}\Delta\mathbf{t}\\0\end{bmatrix}) \tag{35}$$

4.1.1 Attitude Error Vector

First, we examine the attitude error. In our analysis, the vectors representing unit vectors and points in the body frame are denoted by $\mathbf{n}_{\mathbf{b}_i}$ and $\mathbf{p}_{\mathbf{b}_i}$, respectively, while those in the reference frame are represented by $\mathbf{n}_{\mathbf{r}_i}$ and $\mathbf{p}_{\mathbf{r}_i}$. These vectors comprise their true values, which are error-free, plus additional three-dimensional additive errors. Notably, when considering the reference frame as identical to the body frame, the true values of the reference and body vectors are equivalent. This is expressed mathematically as follows:

$$\mathbf{n}_{\mathbf{b}_i} = \mathbf{n}^{\mathbf{t}}_{\mathbf{b}_i} + \Delta\mathbf{n}_{\mathbf{b}_i} \tag{36}$$
$$\mathbf{n}_{\mathbf{r}_i} = \mathbf{n}^{\mathbf{t}}_{\mathbf{b}_i} + \Delta\mathbf{n}_{\mathbf{r}_i} \tag{37}$$
$$\mathbf{p}_{\mathbf{b}_i} = \mathbf{p}^{\mathbf{t}}_{\mathbf{b}_i} + \Delta\mathbf{p}_{\mathbf{b}_i} \tag{38}$$
$$\mathbf{p}_{\mathbf{r}_i} = \mathbf{p}^{\mathbf{t}}_{\mathbf{b}_i} + \Delta\mathbf{p}_{\mathbf{r}_i} \tag{39}$$

In this case, the attitude profile matrix A can be rewritten:

$$\begin{aligned} A &= \frac{1}{2}\left[(C_3^{0^T} + \Delta C_3{}^T)(C_2^0 + C_2{}^T)(C_3^{0^T} + \Delta C_3) - C_1^0 - \Delta C_1 - C_1^{0^T} - \Delta C_1{}^T\right] \\ &= A_0 + \Delta A \end{aligned} \tag{40}$$

The error-free matrix A_0 and the perturbation matrix ΔA are defined as follows:

$$A_0 = 4\begin{bmatrix} B_{0n} + B_{0p} - 2B_0 & \mathbf{0}_{3\times 1} \\ \mathbf{0}_{1\times 3} & k + \sum_{i=1}^{l} \mathbf{p}^{\mathbf{t}}_{\mathbf{b}_i}{}^{\mathbf{T}} \mathbf{p}^{\mathbf{t}}_{\mathbf{b}_i} \end{bmatrix} \tag{41}$$

where

$$B_0 = \frac{1}{l}\sum_{i,j}^{l}\left[[\mathbf{p}^{\mathbf{t}}_{\mathbf{b}_i}\times][\mathbf{p^t}_{\mathbf{b}_j}\times]\right]$$
$$B_{0n} = \sum_{i=1}^{k}\left[[\mathbf{n}^{\mathbf{t}}_{\mathbf{b}_i}\times][\mathbf{n}^{\mathbf{t}}_{\mathbf{b}_i}\times] + \mathbf{n}^{\mathbf{t}}_{\mathbf{b}_i}\mathbf{n}^{\mathbf{t}}_{\mathbf{b}_i}{}^{\mathbf{T}}\right]$$
$$B_{0p} = \sum_{i=1}^{l}\left[[\mathbf{p}^{\mathbf{t}}_{\mathbf{b}_i}\times][\mathbf{p}^{\mathbf{t}}_{\mathbf{b}_i}\times] + \mathbf{p}^{\mathbf{t}}_{\mathbf{b}_i}\mathbf{p}^{\mathbf{t}}_{\mathbf{b}_i}{}^{\mathbf{T}}\right]$$

$$\Delta A = 2\begin{bmatrix} (\Delta B_{0n} + \Delta B_{0p} - 2\Delta B) + (\Delta B_{0n} + \Delta B_{0p} - 2\Delta B)^T & (\Delta Z + \Delta Z_n + \Delta Z_p) \\ (\Delta Z + \Delta Z_n + \Delta Z_p)^T & 0 \end{bmatrix} \tag{42}$$

Manuscript 58 submitted to 64th Israel Annual Conference on Aerospace Sciences (IACAS). Received January 9, 2025.

where

$$\Delta B = \frac{1}{l}\sum_{i,j}^{l}\left[[\mathbf{p}^{\mathbf{t}}_{\mathbf{b_i}}\times]([\mathbf{\Delta p}_{\mathbf{r_j}}\times] + [\mathbf{\Delta p}_{\mathbf{b_j}}\times])\right]$$

$$\Delta B_n = \sum_{i=1}^{k}\left[[\mathbf{n}^{\mathbf{t}}_{\mathbf{b_i}}\times]([\mathbf{\Delta n}_{\mathbf{b_i}}\times] - [\mathbf{\Delta n}_{\mathbf{r_i}}\times]) + \mathbf{\Delta n}_{\mathbf{r_i}}\mathbf{n}^{\mathbf{t}}_{\mathbf{b_i}}{}^{\mathbf{T}} + \mathbf{n}^{\mathbf{t}}_{\mathbf{b_i}}\mathbf{\Delta n}_{\mathbf{b_i}}{}^{\mathbf{T}}\right]$$

$$\Delta B_p = \sum_{i=1}^{l}\left[[\mathbf{p}^{\mathbf{t}}_{\mathbf{b_i}}\times]([\mathbf{\Delta p}_{\mathbf{b_i}}\times] - [\mathbf{\Delta p}_{\mathbf{r_i}}\times]) + \mathbf{\Delta p}_{\mathbf{r_i}}\mathbf{p}^{\mathbf{t}}_{\mathbf{b_i}}{}^{\mathbf{T}} + \mathbf{p}^{\mathbf{t}}_{\mathbf{b_i}}\mathbf{\Delta p}_{\mathbf{b_i}}{}^{\mathbf{T}}\right]$$

$$\Delta Z = \frac{1}{l}\sum_{i,j}^{l}\left[[\mathbf{p}^{\mathbf{t}}_{\mathbf{b_i}}\times](\mathbf{\Delta p}_{\mathbf{r_j}} - \mathbf{\Delta p}_{\mathbf{b_j}})\right]$$

$$\Delta Z_n = \frac{1}{k}\sum_{i=1}^{k}\left[[\mathbf{n}^{\mathbf{t}}_{\mathbf{b_i}}\times](\mathbf{\Delta n}_{\mathbf{b_i}} - \mathbf{\Delta n}_{\mathbf{r_i}})\right]$$

$$\Delta Z_p = \frac{1}{l}\sum_{i=1}^{l}\left[[\mathbf{p}^{\mathbf{t}}_{\mathbf{b_i}}\times](\mathbf{\Delta p}_{\mathbf{b_i}} - \mathbf{\Delta p}_{\mathbf{r_i}})\right]$$

Inserting A_0(Eq. 41) and ΔA(Eq. 42) into the Eq. 26 yields:

$$(A_0 + \Delta A)\left(\begin{bmatrix}\mathbf{0}_3\\1\end{bmatrix} + \frac{1}{2}\begin{bmatrix}\delta\boldsymbol{\theta}\\0\end{bmatrix}\right) = (\lambda_0 + \Delta\lambda)\left(\begin{bmatrix}\mathbf{0}_3\\1\end{bmatrix} + \frac{1}{2}\begin{bmatrix}\delta\boldsymbol{\theta}\\0\end{bmatrix}\right) \tag{43}$$

After performing the multiplication, neglecting second-order terms, and canceling the common zeroth-order term $A_0 I_q = \lambda_0 I_q$ on both sides, the equation simplifies to:

$$\frac{1}{2}A_0\begin{bmatrix}\delta\boldsymbol{\theta}\\0\end{bmatrix} + \Delta A\begin{bmatrix}\mathbf{0}_3\\1\end{bmatrix} = \frac{1}{2}\lambda_0\begin{bmatrix}\delta\boldsymbol{\theta}\\0\end{bmatrix} + \Delta\lambda\begin{bmatrix}\mathbf{0}_3\\1\end{bmatrix} \tag{44}$$

Inserting the explicit forms of A_0 and ΔB yields:

$$\begin{bmatrix}(B_{0n} + B_{0p} - 2B_0)\delta\boldsymbol{\theta} + 2(\Delta Z^T + \Delta Z_n^T + \Delta Z_p^T)\\0\end{bmatrix} = \begin{bmatrix}\frac{1}{2}\lambda_0\delta\boldsymbol{\theta}\\\Delta\lambda\end{bmatrix} \tag{45}$$

The fourth component of the equation confirms that the term is zero to first order with respect to vector errors. The first three components of Eq. (45) are crucial, as they provide the formula for the attitude error vector:

$$\delta\boldsymbol{\theta} = 2\left[\frac{1}{2}\lambda_0 I_3 - (B_{0n} + B_{0p} - 2B_0)\right]^{-1}(\Delta Z + \Delta Z_n + \Delta Z_p) \tag{46}$$

where

$$B_0 = \frac{1}{l}\sum_{i,j}^{l}\left[[\mathbf{p}^{\mathbf{t}}_{\mathbf{b_i}}\times][\mathbf{p}^{\mathbf{t}}{}_{\mathbf{b_j}}\times]\right]$$

$$B_{0n} = \sum_{i=1}^{k}\left[[\mathbf{n}^{\mathbf{t}}_{\mathbf{b_i}}\times][\mathbf{n}^{\mathbf{t}}_{\mathbf{b_i}}\times] + \mathbf{n}^{\mathbf{t}}_{\mathbf{b_i}}\mathbf{n}^{\mathbf{t}}_{\mathbf{b_i}}{}^{\mathbf{T}}\right]$$

$$B_{0p} = \sum_{i=1}^{l}\left[[\mathbf{p}^{\mathbf{t}}_{\mathbf{b_i}}\times][\mathbf{p}^{\mathbf{t}}_{\mathbf{b_i}}\times] + \mathbf{p}^{\mathbf{t}}_{\mathbf{b_i}}\mathbf{p}^{\mathbf{t}}_{\mathbf{b_i}}{}^{\mathbf{T}}\right]$$

$$\Delta Z = \frac{1}{l}\sum_{i,j}^{l}\left[[\mathbf{p}^{\mathbf{t}}_{\mathbf{b_i}}\times](\mathbf{\Delta p}_{\mathbf{r_j}} - \mathbf{\Delta p}_{\mathbf{b_j}})\right]$$

$$\Delta Z_n = \frac{1}{k}\sum_{i=1}^{k}\left[[\mathbf{n}^{\mathbf{t}}_{\mathbf{b_i}}\times](\mathbf{\Delta n}_{\mathbf{b_i}} - \mathbf{\Delta n}_{\mathbf{r_i}})\right]$$

$$\Delta Z_p = \frac{1}{l}\sum_{i=1}^{l}\left[[\mathbf{p}^{\mathbf{t}}_{\mathbf{b_i}}\times](\mathbf{\Delta p}_{\mathbf{b_i}} - \mathbf{\Delta p}_{\mathbf{r_i}})\right]$$

4.1.2 Translation Error Vector

We can now solve for the dual part $\boldsymbol{q_d}$ as a function of the attitude quaternion $\boldsymbol{q}$:

$$\boldsymbol{q_d} = -(C_2 + C_2{}^T)^{-1} C_3{}^T \boldsymbol{q} \tag{47}$$

where $C_3 = C_3^0 + \Delta C_3$ The relationship between the translation vector and the dual quaternion is as follows:

$$\boldsymbol{t} = \frac{1}{2}\begin{bmatrix} \Delta \mathrm{t} \\ 0 \end{bmatrix} = W(\boldsymbol{q})^T \boldsymbol{q_d} \tag{48}$$

From this, we determine the position error vector formula:

$$\mathbf{\Delta t} = \frac{2}{l}\left[\sum_{i=1}^{l} [\mathbf{p}_{\mathbf{b_i}}^{\mathbf{t}} \times]\right] \delta\boldsymbol{\theta} + \sum_{i=1}^{l} (\mathbf{\Delta p_{r_i}} - \mathbf{\Delta p_{b_i}}) \tag{49}$$

4.2 Covariance Analysis

The vector error in the spacecraft's body and reference frames is typically unknown. Therefore, we conduct covariance analysis using statistical distributions with known expected values, assuming that these vector errors have zero mean:

$$E\{\mathbf{\Delta n_{r_i}}\} = \mathbf{E}\{\mathbf{\Delta n_{b_i}}\} = \mathbf{E}\{\mathbf{\Delta p_{r_i}}\} = \mathbf{E}\{\mathbf{\Delta p_{b_i}}\} = \mathbf{0} \tag{50}$$

Typically, the rationale behind this assumption is that any non-zero means of these quantities would be estimated and corrected prior to attitude estimation. In Eq. (46), the only random variables are represented by $\mathbf{\Delta n_{r_i}}, \mathbf{\Delta n_{b_i}}, \mathbf{\Delta p_{r_i}}, \mathbf{\Delta p_{b_i}}$, with all other quantities being deterministic. Since calculating expected values is a linear process, and the cross product of a vector with itself results in zero, it logically follows that:

$$E\{\delta\boldsymbol{\theta}\} = 2\left[\frac{1}{2}\lambda_0 I_3 - (B_{0n} + B_{0p} - 2B_0)\right]^{-1} E\{\Delta Z + \Delta Z_n + \Delta Z_p\} = \mathbf{0_3} \tag{51}$$

Furthermore, we make certain assumptions about the covariance of the vector errors. Assuming that the errors in each vector are uncorrelated, which implies that:

$$E\{\mathbf{\Delta n_{b_i}} \mathbf{\Delta n_{b_j}^T}\} = \mathbf{E}\{\mathbf{\Delta n_{r_i}} \mathbf{\Delta n_{r_j}^T}\} = \mathbf{E}\{\mathbf{\Delta p_{b_i}} \mathbf{\Delta p_{b_j}^T}\} = \mathbf{E}\{\mathbf{\Delta p_{r_i}} \mathbf{\Delta p_{r_j}^T}\} = \mathbf{0} \text{ for } \mathbf{i} \neq \mathbf{j} \tag{52}$$

$$E\{\mathbf{\Delta n_{b_i}} \mathbf{\Delta n_{r_i}^T}\} = \mathbf{E}\{\mathbf{\Delta p_{b_i}} \mathbf{\Delta p_{r_i}^T}\} = \mathbf{E}\{\mathbf{\Delta n_{b_i}} \mathbf{\Delta p_{r_i}^T}\} = \mathbf{E}\{\mathbf{\Delta n_{r_i}} \mathbf{\Delta p_{b_i}^T}\} = \mathbf{0} \text{ for all } \mathbf{i}, \mathbf{j} \tag{53}$$

The measurement covariance matrices defined as follows:

$$\mathrm{R}_{\mathbf{\Delta n_{b_i}}} = \mathrm{E}\{\mathbf{\Delta n_{b_i}} \mathbf{\Delta n_{b_i}^T}\} \tag{54}$$

$$\mathrm{R}_{\mathbf{\Delta n_{r_i}}} = \mathrm{E}\{\mathbf{\Delta n_{r_i}} \mathbf{\Delta n_{r_i}^T}\} \tag{55}$$

$$\mathrm{R}_{\mathbf{\Delta p_{b_i}}} = \mathrm{E}\{\mathbf{\Delta p_{b_i}} \mathbf{\Delta p_{b_i}^T}\} \tag{56}$$

$$\mathrm{R}_{\mathbf{\Delta p_{r_i}}} = \mathrm{E}\{\mathbf{\Delta p_{r_i}} \mathbf{\Delta p_{r_i}^T}\} \tag{57}$$

Our focus is on the variance of the attitude estimates around their mean, which is characterized by the covariance matrix:

$$P_{\theta\theta} = E\{(\delta\boldsymbol{\theta} - E\{\delta\boldsymbol{\theta}\})(\delta\boldsymbol{\theta} - E\{\delta\boldsymbol{\theta}\})^T\} = E\{\delta\boldsymbol{\theta}\delta\boldsymbol{\theta}^T\} \tag{58}$$

Substituting Eq. (46) $\delta\boldsymbol{\theta}$ in Eq. (58) gives:

$$\begin{aligned} P_{\theta\theta} = B(&\sum_{i=1}^{k} [\mathbf{n_{b_i}^t} \times](\mathbf{R}_{\mathbf{\Delta n_{b_i}}} + \mathbf{R}_{\mathbf{\Delta n_{r_i}}})[\mathbf{n_{b_i}^t}{}^{\mathbf{T}} \times] + \sum_{i=1}^{l} [\mathbf{p_{b_i}^t} \times](\mathbf{R}_{\mathbf{\Delta p_{b_i}}} + \mathbf{R}_{\mathbf{\Delta p_{r_i}}})[\mathbf{p_{b_i}^t}{}^{\mathbf{T}} \times] \\ &- \frac{2}{l}\sum_{i,j}^{l} [\mathbf{p_{b_i}^t} \times](\mathbf{R}_{\mathbf{\Delta p_{b_j}}} + \mathbf{R}_{\mathbf{\Delta p_{r_j}}})[\mathbf{p_{b_i}^t}{}^{\mathbf{T}} \times] + \frac{1}{l^2}\sum_{i,j,m}^{l} [\mathbf{p_{b_i}^t} \times](\mathbf{R}_{\mathbf{\Delta p_{b_j}}} + \mathbf{R}_{\mathbf{\Delta p_{r_j}}})[\mathbf{p_{b_m}^t}{}^{\mathbf{T}} \times])\mathbf{B} \end{aligned} \tag{59}$$

where

$$B = 2[\frac{1}{2}\lambda_0 I_3 - (B_{0n} + B_{0p} - 2B_0)]^{-1}$$

Manuscript 58 submitted to 64th Israel Annual Conference on
Aerospace Sciences (IACAS). Received January 9, 2025.

and λ_0 is the max eigenvalue of the A_0 matrix. The translation error vector is unbiased since it's linear to the $\boldsymbol{\delta\theta}$:

$$E\{\boldsymbol{\Delta t}\} = E\{\frac{2}{l}\left[\sum_{i=1}^{l}[\mathbf{p}^{\mathbf{t}}_{\mathbf{b}_i\times}]\right]\boldsymbol{\delta\theta} + \sum_{i=1}^{l}(\boldsymbol{\Delta}\mathbf{p}_{\mathbf{r}_i} - \boldsymbol{\Delta}\mathbf{p}_{\mathbf{b}_i})\} = \mathbf{0}_3 \tag{60}$$

Then, the covariance of the translation error vector is as follows:

$$\begin{aligned} P_{tt} = &\frac{4}{l^2}[\sum_{i=1}^{l}[\mathbf{p}^{\mathbf{t}}_{\mathbf{b}_i}\times]P_{\theta\theta}[\mathbf{p}^{\mathbf{t}}_{\mathbf{b}_i}{}^{\mathrm{T}}\times] + \sum_{i=1}^{l}(\mathbf{R}_{\boldsymbol{\Delta}\mathbf{p}_{\mathbf{b}_i}} + \mathbf{R}_{\boldsymbol{\Delta}\mathbf{p}_{\mathbf{r}_i}}) \\ &+ 2B(\frac{1}{l}\sum_{i,j,m}^{l}[\mathbf{p}^{\mathbf{t}}_{\mathbf{b}_i}\times][\mathbf{p}^{\mathbf{t}}_{\mathbf{b}_j}\times](\mathbf{R}_{\boldsymbol{\Delta}\mathbf{p}_{\mathbf{r}\mathbf{m}}} + \mathbf{R}_{\boldsymbol{\Delta}\mathbf{p}_{\mathbf{b}\mathbf{m}}}) - \sum_{i,j}^{l}[\mathbf{p}^{\mathbf{t}}_{\mathbf{b}_i}\times][\mathbf{p}^{\mathbf{t}}_{\mathbf{b}_j}\times](\mathbf{R}_{\boldsymbol{\Delta}\mathbf{p}_{\mathbf{r}_j}} + \mathbf{R}_{\boldsymbol{\Delta}\mathbf{p}_{\mathbf{b}_j}}))] \end{aligned} \tag{61}$$

Summary of Error and Covariance Expressions

Under the assumption that the body frame and reference frame coincide, given the three-dimensional point observations, $\{\mathbf{p}_{\mathbf{b}_i}\}_{i=1}^{l}$, and unit vector observations, $\{\mathbf{n}_{\mathbf{b}_i}\}_{i=1}^{k}$, which are equal to their true values($\{\mathbf{p}^{\mathbf{t}}_{\mathbf{b}_i}\}_{i=1}^{l}$, $\{\mathbf{n}^{\mathbf{t}}_{\mathbf{b}_i}\}_{i=1}^{k}$) plus the three-dimensional additive errors. Compute the following expressions:

$$\Delta Z = \frac{1}{l}\sum_{i,j}^{l}\left[[\mathbf{p}^{\mathbf{t}}_{\mathbf{b}_i}\times](\boldsymbol{\Delta}\mathbf{p}_{\mathbf{r}_j} - \boldsymbol{\Delta}\mathbf{p}_{\mathbf{b}_j})\right]$$

$$\Delta Z_n = \frac{1}{k}\sum_{i=1}^{k}\left[[\mathbf{n}^{\mathbf{t}}_{\mathbf{b}_i}\times](\boldsymbol{\Delta}\mathbf{n}_{\mathbf{b}_i} - \boldsymbol{\Delta}\mathbf{n}_{\mathbf{r}_i})\right]$$

$$\Delta Z_p = \frac{1}{l}\sum_{i=1}^{l}\left[[\mathbf{p}^{\mathbf{t}}_{\mathbf{b}_i}\times](\boldsymbol{\Delta}\mathbf{p}_{\mathbf{b}_i} - \boldsymbol{\Delta}\mathbf{p}_{\mathbf{r}_i})\right]$$

The expressions of the three-dimensional Euler vector error and translation error vector yield:

$$\boldsymbol{\delta\theta} = 2\left[\frac{1}{2}\lambda_0 I_3 - (B_{0n} + B_{0p} - 2B_0)\right]^{-1}(\Delta Z + \Delta Z_n + \Delta Z_p) \tag{62}$$

$$\boldsymbol{\Delta t} = \frac{2}{l}\left[\sum_{i=1}^{l}[\mathbf{p}^{\mathbf{t}}_{\mathbf{b}_i\times}]\right]\boldsymbol{\delta\theta} + \sum_{i=1}^{l}(\boldsymbol{\Delta}\mathbf{p}_{\mathbf{r}_i} - \boldsymbol{\Delta}\mathbf{p}_{\mathbf{b}_i}) \tag{63}$$

Then, a random error analysis for their covariance matrices. 3×3 covariance matrices of the rotation error and translation error are developed in the following expressions:

$$\begin{aligned} P_{\theta\theta} = &B(\sum_{i=1}^{k}[\mathbf{n}^{\mathbf{t}}_{\mathbf{b}_i}\times](\mathbf{R}_{\boldsymbol{\Delta}\mathbf{n}_{\mathbf{b}_i}} + \mathbf{R}_{\boldsymbol{\Delta}\mathbf{n}_{\mathbf{r}_i}})[\mathbf{n}^{\mathbf{t}}_{\mathbf{b}_i}{}^{\mathrm{T}}\times] + \sum_{i=1}^{l}[\mathbf{p}^{\mathbf{t}}_{\mathbf{b}_i}\times](\mathbf{R}_{\boldsymbol{\Delta}\mathbf{p}_{\mathbf{b}_i}} + \mathbf{R}_{\boldsymbol{\Delta}\mathbf{p}_{\mathbf{r}_i}})[\mathbf{p}^{\mathbf{t}}_{\mathbf{b}_i}{}^{\mathrm{T}}\times] \\ &- \frac{2}{l}\sum_{i,j}^{l}[\mathbf{p}^{\mathbf{t}}_{\mathbf{b}_i}\times](\mathbf{R}_{\boldsymbol{\Delta}\mathbf{p}_{\mathbf{b}_j}} + \mathbf{R}_{\boldsymbol{\Delta}\mathbf{p}_{\mathbf{r}_j}})[\mathbf{p}^{\mathbf{t}}_{\mathbf{b}_i}{}^{\mathrm{T}}\times] + \frac{1}{l^2}\sum_{i,j,m}^{l}[\mathbf{p}^{\mathbf{t}}_{\mathbf{b}_i}\times](\mathbf{R}_{\boldsymbol{\Delta}\mathbf{p}_{\mathbf{b}_j}} + \mathbf{R}_{\boldsymbol{\Delta}\mathbf{p}_{\mathbf{r}_j}})[\mathbf{p}^{\mathbf{t}}_{\mathbf{b}_m}{}^{\mathrm{T}}\times])\mathbf{B} \end{aligned} \tag{64}$$

$$\begin{aligned} P_{tt} = &\frac{4}{l^2}[\sum_{i=1}^{l}[\mathbf{p}^{\mathbf{t}}_{\mathbf{b}_i}\times]P_{\theta\theta}[\mathbf{p}^{\mathbf{t}}_{\mathbf{b}_i}{}^{\mathrm{T}}\times] + \sum_{i=1}^{l}(\mathbf{R}_{\boldsymbol{\Delta}\mathbf{p}_{\mathbf{b}_i}} + \mathbf{R}_{\boldsymbol{\Delta}\mathbf{p}_{\mathbf{r}_i}}) \\ &+ 2B(\frac{1}{l}\sum_{i,j,m}^{l}[\mathbf{p}^{\mathbf{t}}_{\mathbf{b}_i}\times][\mathbf{p}^{\mathbf{t}}_{\mathbf{b}_j}\times](\mathbf{R}_{\boldsymbol{\Delta}\mathbf{p}_{\mathbf{r}\mathbf{m}}} + \mathbf{R}_{\boldsymbol{\Delta}\mathbf{p}_{\mathbf{b}\mathbf{m}}}) - \sum_{i,j}^{l}[\mathbf{p}^{\mathbf{t}}_{\mathbf{b}_i}\times][\mathbf{p}^{\mathbf{t}}_{\mathbf{b}_j}\times](\mathbf{R}_{\boldsymbol{\Delta}\mathbf{p}_{\mathbf{r}_j}} + \mathbf{R}_{\boldsymbol{\Delta}\mathbf{p}_{\mathbf{b}_j}}))] \end{aligned} \tag{65}$$

where

$$B = 2[\frac{1}{2}\lambda_0 I_3 - (B_{0n} + B_{0p} - 2B_0)]^{-1}$$

Equations. (64)-(65) *constitute the main results of this work.* It is the first time that expressions for the covariance matrices of the estimation errors of this dual quaternion estimator. The availability of these matrices means that the practitioners can assess the accuracy of the pose estimates, which is arguably as important as the estimated values themselves.

Manuscript 58 submitted to 64th Israel Annual Conference on
Aerospace Sciences (IACAS). Received January 9, 2025.

5. Simulation

This section validates the consistency of rotation error, translation error, and their covariance expressions through Monte Carlo simulations in specific scenarios. It also evaluates the algorithm's performance across varying numbers of input datasets and different weighting schemes.

5.1 Statistical Consistency

Figure. 2 is the schematic diagram of coordinate system rotation and translation:

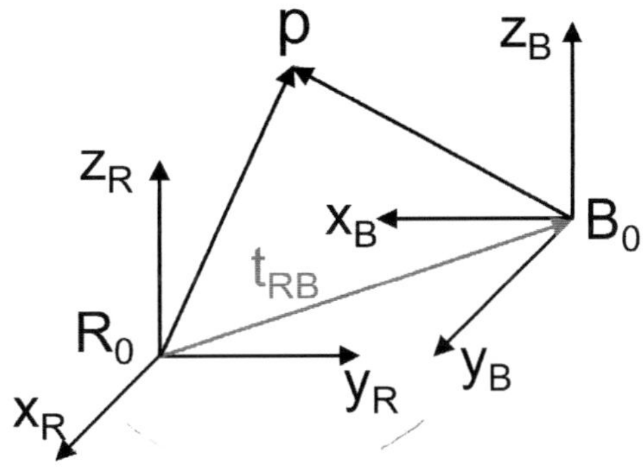

Figure 2: Schematics of the Relative Pose between Frames

To verify the consistency of the predicted errors and covariance, specific values for the attitude quaternion, translation vector, and the measurements are set as follows:

$$
\begin{aligned}
\boldsymbol{q}_{RB} &= \begin{bmatrix} 0 & 0 & -\frac{\sqrt{2}}{2} & \frac{\sqrt{2}}{2} \end{bmatrix} \\
\mathbf{t}_{RB} &= \begin{bmatrix} 0 & 0 & 2 \end{bmatrix} \\
\mathbf{n}_{\mathbf{r}_1} &= \begin{bmatrix} 1 & 0 & 0 \end{bmatrix} \\
\mathbf{n}_{\mathbf{r}_2} &= \begin{bmatrix} 0 & 1 & 0 \end{bmatrix} \\
\mathbf{p}_{\mathbf{r}_1} &= \begin{bmatrix} -2 & 2 & 0 \end{bmatrix} \\
\mathbf{p}_{\mathbf{r}_2} &= \begin{bmatrix} 2 & 0 & 2 \end{bmatrix}
\end{aligned}
$$

A sample of 100,000 runs was created, and the Monte Carlo (MC) values were calculated using the standard sample average. Unit vectors and points in the reference coordinate system are transformed into the body coordinate system by Eqs. (14),(15). Assume that the additional noise $\Delta\mathbf{n}_{\mathbf{r}_i}$, $\Delta\mathbf{n}_{\mathbf{b}_i}$, $\Delta\mathbf{p}_{\mathbf{r}_i}$, $\Delta\mathbf{p}_{\mathbf{b}_i}$ have the same standard deviation $\sigma = 0.01$ and their covariance matrices are diagonal yields.

$$
\begin{aligned}
\mathrm{R}_{\Delta\mathbf{n}_{\mathbf{b}_i}} &= \sigma^2 \mathrm{I}_3 \\
\mathrm{R}_{\Delta\mathbf{n}_{\mathbf{r}_i}} &= \sigma^2 \mathrm{I}_3 \\
\mathrm{R}_{\Delta\mathbf{p}_{\mathbf{b}_i}} &= \sigma^2 \mathrm{I}_3 \\
\mathrm{R}_{\Delta\mathbf{p}_{\mathbf{r}_i}} &= \sigma^2 \mathrm{I}_3
\end{aligned}
$$

Table. 1 summarizes the 3D angular and translation error results. In this case, the Monte Carlo test results and predicted values are both 10^{-5}.

Manuscript 58 submitted to 64th Israel Annual Conference on
Aerospace Sciences (IACAS). Received January 9, 2025.

Table 1: Statistical Consistency of the Error Vector

$\delta\theta^{MC}(\times10^{-4})$	$\delta\theta(\times10^{-4})$	$\Delta t^{MC}(\times10^{-4})$	$\Delta t(\times10^{-4})$
$\begin{bmatrix} -0.2905 \\ 0.0454 \\ 0.0657 \end{bmatrix}$	$\begin{bmatrix} -0.0813 \\ -0.0045 \\ 0.1755 \end{bmatrix}$	$\begin{bmatrix} 0.4273 \\ -0.3374 \\ 2.5170 \end{bmatrix}$	$\begin{bmatrix} -0.2248 \\ 1.1247 \\ -0.9851 \end{bmatrix}$

Tables. 2, 3 summarize the covariance matrices results for the attitude and translation errors. The results show that the formula predictions for both attitude error and translation error are on the same order of magnitude as the Monte Carlo results, approximately σ^2. The magnitudes and directions implied by the eigenvalues between the theoretical predictions and the Monte Carlo covariance matrix results match well, validating the accuracy of the theoretical model in predicting how the errors behave in real-world scenarios.

Table 2: Covariance Results of Attitude Error Vector

${P_{\delta\theta}}^{MC}(\times10^{-3})$	$P_{\delta\theta}(\times10^{-3})$	eigenvalue of ${P_{\delta\theta}}^{MC}, P_{\delta\theta}(\times10^{-3})$
$\begin{bmatrix} 0.1415 & -0.0453 & 0.0366 \\ -0.0453 & 0.0726 & -0.0181 \\ 0.0366 & -0.0181 & 0.0545 \end{bmatrix}$	$\begin{bmatrix} 0.2561 & -0.1277 & 0.0685 \\ -0.1277 & 0.1121 & -0.0479 \\ 0.0685 & -0.0479 & 0.0528 \end{bmatrix}$	$\begin{bmatrix} 0.0409, 0.3539 \\ 0.0501, 0.0412 \\ 0.1776, 0.0260 \end{bmatrix}$

Table 3: Covariance Results for the Translation Error Vector

${P_{\Delta t}}^{MC}(\times10^{-3})$	$P_{\Delta t}(\times10^{-3})$	eigenvalue of ${P_{\Delta t}}^{MC}, P_{\Delta t}(\times10^{-3})$
$\begin{bmatrix} 0.5249 & 0.1048 & -0.1051 \\ 0.1048 & 0.5240 & -0.1243 \\ -0.1051 & -0.1243 & 0.5221 \end{bmatrix}$	$\begin{bmatrix} 1.1287 & 0.3691 & 0.4307 \\ 0.3691 & 0.8065 & 0.1176 \\ 0.4307 & 0.1176 & 0.7704 \end{bmatrix}$	$\begin{bmatrix} 0.3987, 1.5888 \\ 0.4255, 0.4565 \\ 0.7468, 0.6602 \end{bmatrix}$

Using the same attitude quaternion and two reference vectors, the attitude covariance matrix and Monte Carlo test results based on the q-method are shown in Tab. 4, which proves again the accuracy of the derived formula:

Table 4: Covariance Results of q-method

${P_{\delta\theta}}^{MC}(\times10^{-3})$	$P_{\delta\theta}(\times10^{-3})$	eigenvalue of ${P_{\delta\theta}}^{MC}, P_{\delta\theta}(\times10^{-3})$
$\begin{bmatrix} 0.0993 & 0.0001 & 0.0001 \\ 0.0001 & 0.2001 & 0.0000 \\ 0.0001 & 0.0000 & 0.0705 \end{bmatrix}$	$\begin{bmatrix} 0.2000 & 0.0000 & 0.0000 \\ 0.0000 & 0.2000 & 0.0000 \\ 0.0000 & 0.0000 & 0.1000 \end{bmatrix}$	$\begin{bmatrix} 0.0705, 0.1000 \\ 0.0993, 0.2000 \\ 0.2001, 0.2000 \end{bmatrix}$

5.2 Performance Evaluation

To evaluate the accuracy of our dual quaternion estimation algorithm, we executed an extensive simulation involving 100,000 trials. Dual quaternions were randomly generated during these trials with rotation quaternion components following a Gaussian distribution of zero mean and unit covariance, then normalized for validity. The translation components of the dual quaternions were derived from their corresponding rotation quaternions, ensuring they properly represent translations as per dual quaternion formulation. Each element of the 3D translation vector was uniformly distributed between 0 and 10 meters.

An equal number of k-unit vectors and points were used for our simulation, distributed randomly over a 100x100x100 meter cubic volume. Each set of dual quaternions generated k random reference unit vectors and points across the cube's surface, which were then transformed into the body frame using the respective dual quaternion. Gaussian noise was added to both data sets: unit vectors had a noise standard deviation of 1 degree per axis, and points had 0.5 cm per axis.

These body data and reference data are used to calculate the estimated rotation and translation parameters. The angle error $\delta\theta$ and the translation error $\mathbf{\Delta t}$ are calculated as follows:

$$\delta\theta = 2\arccos(\boldsymbol{q}^T \boldsymbol{q}) \tag{66}$$

$$\Delta t = |\hat{\mathbf{t}} - \mathbf{t}| \tag{67}$$

Manuscript 58 submitted to 64th Israel Annual Conference on Aerospace Sciences (IACAS). Received January 9, 2025.

Figure. 3 presents the cumulative distribution function (CDF) plots for angular and translation errors to evaluate the performance of the pose estimation algorithm under three different weighting configurations for 2 unit vectors and 2 points. The weight ratios of the unit vector and point of the three configurations are 10000, 1, and 0.0001, respectively. The left graph illustrates the angular error ($\delta\theta$) in degrees, while the right graph presents the translation error (Δt) in meters.
The angular error CDF plot reveals that Configuration 1($\alpha = 10000, \beta = 1$), which allocates a higher weight to unit vectors, registers the poorest performance among the three tested configurations. This indicates that an over-reliance on unit vectors is detrimental to achieving the best angle and position accuracy in this particular setting. Intriguingly, despite its not optimal results, Configuration 1 still surpasses the performance of the traditional q-method pose estimation algorithm, which exclusively relies on unit vectors. In addition, the rotational and translational errors of configuration 2($\alpha = 1, \beta = 1$) and configuration 3($\alpha = 1, \beta = 10000$) are very close, reflecting the benefits of a more balanced weighting approach.
Regarding translation errors, configurations 2 and 3 perform better than configuration 1, confirming the key role of point measurements in ensuring accurate rotation and translation estimates. The performances of the equal-weighted and high-point-weighted configurations are closely matched, underscoring the significant influence of appropriately balanced point data on the effectiveness of the pose estimation process. The broader error distributions observed in Fig. 3, due to the limited number of input data (n=2, p=2), suggest that a scant dataset can compromise the reliability of angle and translation estimations.

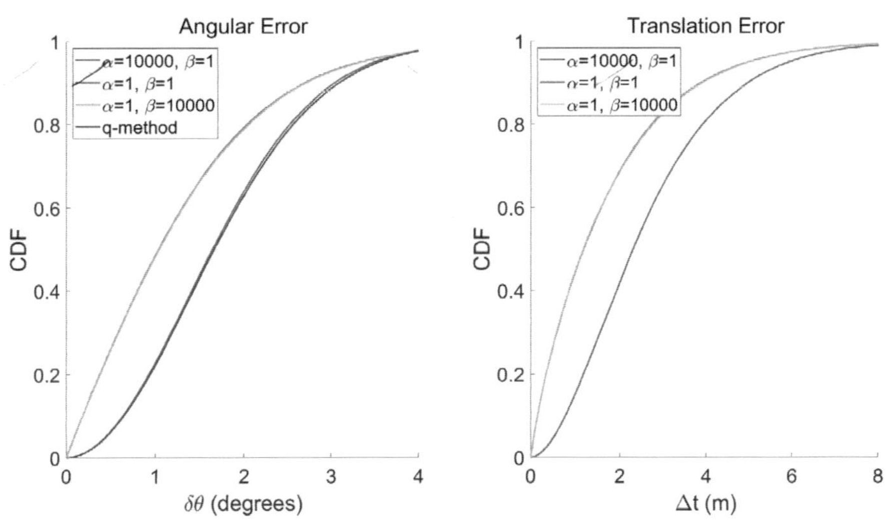

Figure 3: Pose Errors Distributions for Various Weights in the Cost Function

We also discuss the algorithm's performance on different input dataset sizes. Fig. 4 displays the cumulative distribution functions (CDF) of the angular and translation errors derived from simulations of the dual quaternion estimation algorithm, delineating the impact of varying sample sizes. The trends for both angular and translation errors are consistent. As the sample size increases, there is a marked improvement in accuracy. Hence, employing more points and vectors substantially enhances the accuracy of both angle and translation measurements, reflecting the algorithm's capability to optimize pose estimation as the data richness increases. Notably, the CDF curve improves slowly when the number of data points and vectors exceeds 10. Although increasing input data enhances estimation accuracy, it also raises the computational load. Therefore, a balance must be struck between computational speed and estimation accuracy, especially in scenarios where real-time processing is crucial.

Manuscript 58 submitted to 64th Israel Annual Conference on Aerospace Sciences (IACAS). Received January 9, 2025.

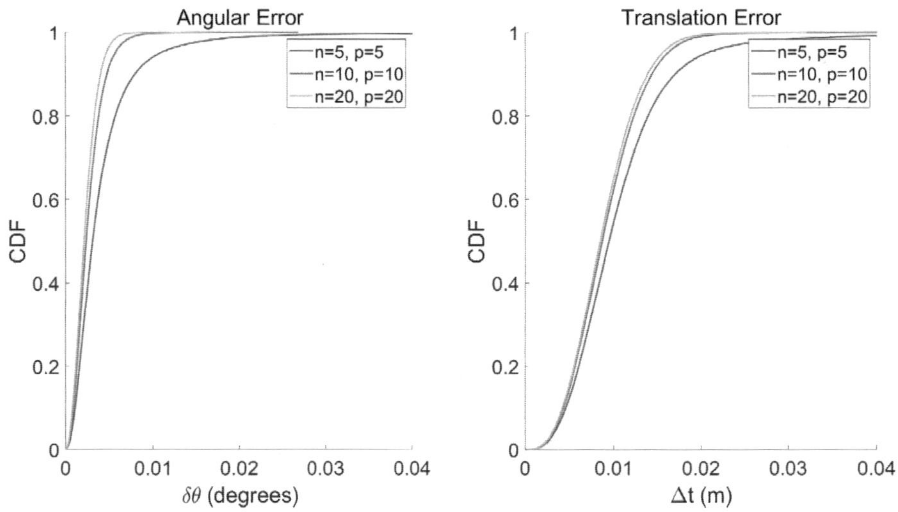

Figure 4: Pose errors Distributions for Various Batch Sizes

6. Conclusions

This study focuses on the error analysis of a dual quaternion estimator that processes batches of point and unit vector observations. The analysis approximates the pose errors to first order in the sensing errors and develops expressions for the covariance matrices of the three-dimensional angular and translational error vectors, which offers deeper insights into how measurement noise and model affect the estimated dual quaternion. These error metrics are essential for evaluating the estimator's accuracy and robustness in practical applications. Extensive Monte Carlo simulations verified the proposed error analysis. To validate the proposed method, extensive Monte Carlo simulations are conducted, comparing the estimated results against the true values across various scenarios. These simulations confirm the accuracy of the derived error expressions and demonstrate the reliability of the estimator under different noise conditions. The performance sensitivity to the batch size and to the weights in the cost function was numerically studied. Larger batch sizes improve robustness and reduce estimation error, while cost weights allow fine-tuning of the balance between rotation and translation contributions in the optimization process.

References

[1] W.R. Hamilton. "On Quaternions; or on a New System of Imaginaries in Algebra". In: *The London, Edinburgh, and Dublin Philosophical Magazine and Journal of Science* 25.169 (1844), pp. 489–495.

[2] W.K. Clifford. "Preliminary Sketch of Biquaternions". In: *Proceedings of the London Mathematical Society* 1.1 (1871), pp. 381–395.

[3] I. Fischer. *Dual-Number Methods in Kinematics, Statics and Dynamics*. Routledge, 2017.

[4] A.T. Yang and F. Freudenstein. "Application of Dual-Number Quaternion Algebra to the Analysis of Spatial Mechanisms". In: *Journal of Applied Mechanics* 31.2 (1964), pp. 300–308.

[5] O. Bottema and B. Roth. *Theoretical Kinematics*. Courier Corporation, 1990.

[6] J.R. Dooley and J.M. McCarthy. "Spatial Rigid Body Dynamics using Dual Quaternion Components". In: *IEEE International Conference on Robotics and Automation*. 1991, pp. 90–95.

[7] V. Brodsky and M. Shoham. "Dual Numbers Representation of Rigid Body Dynamics". In: *Mechanism and Machine Theory* 34.5 (1999), pp. 693–718.

[8] J. Angeles. "The Application of Dual Algebra to Kinematic Analysis". In: *Computational Methods in Mechanical Systems: Mechanism Analysis, Synthesis, and Optimization*. 1998, pp. 3–32.

[9] K. Daniilidis. "Hand-Eye Calibration using Dual Quaternions". In: *The International Journal of Robotics Research* 18.3 (1999), pp. 286–298.

[10] K.S. Arun, T.S. Huang, and S.D. Blostein. "Least-Squares Fitting of Two 3-D Point Sets". In: *IEEE Transactions on Pattern Analysis and Machine Intelligence* 5 (1987), pp. 698–700.

[11] W. Rutkowski, R. Benton, and E. Kent. "Model-Driven Determination of Object Pose for a Visually Servoed Robot". In: *IEEE International Conference on Robotics and Automation*. Vol. 4. 1987, pp. 1419–1428.

[12] O.D. Faugeras and M. Hebert. "A 3-D Recognition and Positioning Algorithm using Geometrical Matching between Primitive Surfaces". In: *Proceedings of the Eighth International Joint Conference on Artificial Intelligence*. Vol. 2. 1983, pp. 996–1002.

[13] M.W. Walker, L. Shao, and R.A. Volz. "Estimating 3-D Location Parameters using Dual Number Quaternions". In: *CVGIP: Image Understanding* 54.3 (1991), pp. 358–367.

[14] Clifford. "Preliminary sketch of biquaternions". In: *Proceedings of the London Mathematical Society* 1.1 (1871), pp. 381–395.

[15] Michel Chasles. "Note sur les propriétés génerales du système de deux corps semblables entr'eux et placés d'une manière quelconque dans l'espace; et sur le déplacement fini ou infiniment petit d'un corps solide libre". In: *Bulletin des Sciences Mathématiques, Férussac* 14 (1830), pp. 321–26.

Midcourse Guidance for Hypersonic Glide Vehicle Interception

Michael Urman *, Oded Golan †, and Vitaly Shaferman ‡
Technion, Israel Institute of Technology, Haifa, Israel, 3200003

Abstract

Hypersonic Glide Vehicles (HGVs) pose a substantial military threat. They have a wide range, high maneuverability, and can fly at lower altitudes compared to ballistic missiles, making it difficult for ground radars to detect them. Hence, the interception of HGVs poses a difficult challenge to air defense systems. The objective of this work is to develop a strategy for HGV interception. The HGV attacker tries to reach a ground target while performing a lateral divert evasive maneuver. An optimal control based interception strategy is developed using a one-sided approach, assuming knowledge of the HGV's target. The strategy aims to maximize the capture zone while accounting for the uncertainty in the HGV's trajectory.

To design the interceptor's strategy, it is essential to investigate the HGV capabilities which can be done by finding the HGV's optimal trajectories, considering the dynamics and the constraints. Two extreme cases of the HGV's behavior are considered: the maximum divert trajectory, where the HGV achieves maximum lateral deviation, and the immediate divert trajectory, where the HGV turns towards its target immediately after the engagement starts. The interceptor aims to a virtual target which follows an intermediate trajectory which is defined between these two extreme trajectories. As the engagement progresses, the immediate trajectory and the virtual target are continuously updated. As long as the HGV's future path remains uncertain, the interceptor aims at the virtual target. Once the HGV turns toward its target, its trajectory is determined, and the interceptor switches to engage the real target.

The interceptor's guidance law follows an optimal trajectory from its current state to the interception point. Every time the intermediate trajectory is updated, an optimal control problem is solved to intercept the updated virtual target. Between updates, the interceptor is guided using the control inputs obtained from the most recent solution. To simplify the algorithm, a modified guidance law is proposed in which the expected intercept point on each virtual trajectory is used to form a time-varying Predicted Intercept Point (PIP). Thus, instead of relying on individual trajectories calculated in advance, we base the guidance law on a time-varying PIP curve.

Both the HGV's and interceptor's trajectory optimization problems are formulated and solved in 3D using a hp-adaptive Pseudospectral method with the Legendre-Gauss-Radau (LGR) collocation approach. The optimization assumes a non-rotating spherical Earth model and focuses on the gliding phase of the HGV. The proposed guidance laws (i.e., trajectory-based and PIP-based) are evaluated in simulation and demonstrate successful interception, with

*Graduate student, Faculty of Aerospace Engineering, Technion, michaelur@campus.technion.ac.il
†Full Research Fellow, Faculty of Aerospace Engineering, Technion, oded.golan@technion.ac.il
‡Associate Professor, Faculty of Aerospace Engineering, Technion, vitalysh@technion.ac.il

Manuscript 60 submitted to 64th Israel Annual Conference on Aerospace Sciences (IACAS). Received January 9, 2025.

similar performance, despite the uncertainties in the HGV's trajectory.

Ablation Computational Simulation using Ansys Fluent

Over the last few years, the interest in Ablative processes has increased dramatically. Among other uses, two common ablative processes are thermal protection and solid rocket motor (SRM). The interest in these uses is linked to the increasing number of projects around high speed hypersonic flying vehicles.

External ablative coating is commonly used for flying vessels that fly at hypersonic velocities, re-entering the atmosphere etc. The ablation process for Solid Rocket Motors is connected to the combustion of solid fuels, where the geometry of the solid fuel can be complex and varied, the recession of the solid fuel surfaces needs to be calculated using computational simulations.

The ablation simulation in Ansys Fluent allows the use of Vielle's model and a surface reaction model to calculate the recession combined with integrated dynamic mesh options to deform the mesh accordingly. These options can include within the Computational Fluid Dynamics calculation except from the pressure and velocity fields also species transport, combustion, volumetric chemical reactions and heat and mass transfer for the fluid and solid, allowing the user to accurately simulate the entire ablative process in a simple and efficient user interface, all in Fluent as a "one stop shop". The Ablation workflow in Fluent is a new and exciting feature as it allows the engineer to define a complete, complex process with minimum use of user defined functions and robust connection between the different components of the calculation

The presentation will include an example for a common workflow and show a few benchmarks for ablating processes calculations including results validation vs. experimental measurements.

Manuscript 48 submitted to 64th Israel Annual Conference on Aerospace Sciences (IACAS). Received January 7, 2025.

Abstract for the 64-th IACAS 2025 Meeting

Flow simulation of waveriders at off-design conditions

Angelos Klothakis*

Technical University of Crete, Chania 73100, Greece

Ioannis K. Nikolos†

Technical University of Crete, Chania 73100, Greece

Vassilis Theofilis‡

Technion - Israel Institute of Technology, Haifa, Israel

Flight at hypersonic speeds introduces severe flow conditions, strong aerodynamic heating, and complex shockwave-boundary layer interactions. Designing vehicles capable of flying under such conditions requires both robust design methodologies and advanced simulation tools capable of accurately predicting flow behavior across a wide range of altitudes and ambient pressures. Among the various hypersonic glide vehicle concepts, waveriders—originally proposed independently in the UK by Nonweiler[1] and the USSR by Maikapar[2]—stand out due to their characteristic shockwaves attached at the leading edges. This configuration effectively suppresses pressure spillage between the pressure and suction sides, thereby maximizing lift and minimizing drag in hypersonic flight.

Our team has designed a waverider intended for Mach-7 flight at altitudes around 90 km[3] using a design approach based on osculating cone theory[4] and validated by DSMC (Direct Simulation Monte Carlo) simulations, with the open-source solver SPARTA.[5] The resulting geometry was optimized for aerodynamic efficiency under these high-altitude conditions. We subsequently tested the vehicle in two off-design flight scenarios: (i) 45 km altitude, Mach 6.2, and 0 degrees angle-of-attack, and (ii) 30 km altitude, Mach 6, and 2 degrees angle-of-attack. Both cases were selected to capture realistic deviations in flight parameters, such as slight angle-of-attack adjustments, and were investigated using the SU2 solver.[6] Figure 1, for instance, presents pressure contours around the vehicle at the 45 km condition.

*Research Scientist, School of Production Engineering and Management, TUC

†Professor, School of Production Engineering and Management, TUC

‡Professor, Departement of Aerospace Engineering, Technion

Manuscript 66 submitted to 64th Israel Annual Conference on Aerospace Sciences (IACAS). Received January 9, 2025.

Parallel efforts to predict laminar-to-turbulent transition at Mach numbers ranging from $4.5 \leq Ma \leq 9$, facilitated by the DSMC method, further demonstrated the influence of wall-slip velocity and temperature-jump conditions on the classical boundary layer stability predictions.[7]

In this work, we extend our earlier research to analyze the vehicle's performance at lower altitudes and higher angles of attack, focusing on whether the original flow structures—including the shockwave attached to the leading edges and the detached conical shock—remain unchanged. We also assess the surface temperature distributions arising in denser atmospheric conditions and quantify the Lift-to-Drag ratio, a crucial metric for evaluating whether the waverider can maintain adequate flying capabilities under markedly off-design conditions. By examining the aerothermal environment and aerodynamic performance in these more demanding flight regimes, our study aims to enhance the design robustness and operational flexibility of hypersonic waveriders. Detailed findings will be discussed at the time of the conference.

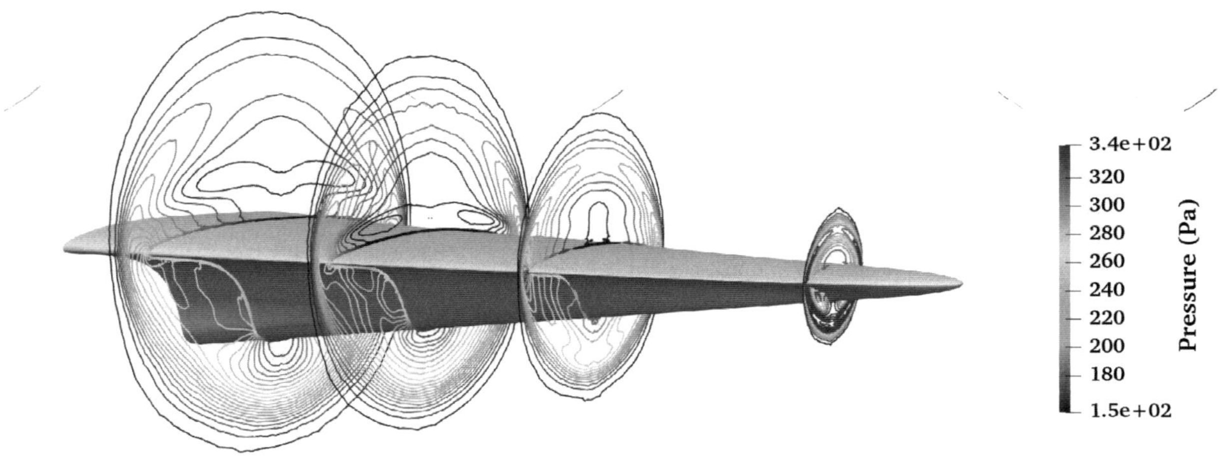

Figure 1: Pressure contours around the waverider computed in[8] at an altitude of 45 km

References

[1]Nonweiler, T. R. F., "Aerodynamic Problems of Manned Space Vehicles," *Journal of the Royal Aeronautical Society*, Vol. 63, 1959, pp. 521–528.

[2]Maikapar, G. I., "On the wave drag of non-axisymmetric bodies at supersonic speeds," *Journal of Applied Mathematics and Mechanics*, Vol. 23, No. 2, 1959, pp. 528–531.

[3]Klothakis, A. and Nikolos, I. K., "Evaluation of a Hypersonic Waverider Vehicle Using DSMC," *Computation*, Vol. 12, 2024, pp. 148.

[4]Sobieczky, H., Dougherty, F., and Jones, K. D., "Hypersonic Waverider Design from Given Shock Waves," *First International Hypersonic Waverider Symposium, Univ. of Maryland, College Park, MD*, 1990.

[5]Gallis, M. A., Torczynski, J. R., Plimpton, S. J., Rader, D. J., and Koehler, T., "Direct Simulation Monte Carlo: The quest for speed," *29th Rarefied Gas Dynamics (RGD) Symposium, Xi'an, China*, 2014.

[6]Economon, T., Palacios, F., Copeland, S., Lukaczyk, T., and Alonso, J. J., "SU2: An open-source suite for multiphysics simulation and design," *AIAA Journal*, Vol. 54, No. 3, 2016, pp. 828–846.

[7]Klothakis, A., Quintanilha, H., Sawant, S. S., Protopapadakis, E., Theofilis, V., and Levin, D. A., "Linear stability analysis of hypersonic boundary layers computed by a kinetic approach: a semi-infinite flat plate at $4.5 \leq Ma \leq 9$," *Theor. Comput. Fluid Dyn.*, Vol. 36, 2022, pp. 117–139.

[8]Klothakis, A. and Nikolos, I. K., "Simulation of a Mach-7 waverider in off-design conditions," *3rd Integrated Air and Missile Defense Center of Excellence Annual Conference, Chania, Greece*, 2024.

Wall Function LES using Fluent GPU solver and Fluent Rapid Octree mesh

Over the last few years hardware technology advancements opened new possibilities for high fidelity simulation of large and complex models. While RANS simulations are very efficient in terms of resources (using steady state solutions for example) and can predict quite well aerodynamic coefficients at the "linear part" of a polar sweep, their results are consistently far at the stall region.

Using the Large Eddie Simulation (LES) method engineers can produce high fidelity results also for the stall region, using large meshes and adequate hardware. The Wall Function LES (also known as WFLES or WMLES - Wall Model LES) approach can produce high fidelity solution using wall functions to consider boundary layer effects. That allows the use of low aspect ratio mesh cells near the walls which make the Rapid Octree (RO) meshing method a great fit.

RO is an octree, hexahedral, top to bottom meshing approach, which parallelizes very well. That allows the production of very large and high quality meshes extremely fast. Fluent RO allows a neat workflow for meshing which is very easy to use and user friendly.

The presentation will include an example of a 1 billion cells mesh, produced in less than 30 minutes on 512 CPU cores. The calculation time was 1 hour per Convective Time Unit (CTU) on 32 L40 GPU's. Simulation results showed good agreement with the experimental results in the stall region.

Overall, the presentation will show an exciting workflow that leverages the most recent hardware technology advancements by using new and efficient meshing approach and high fidelity solver that runs on GPU, while giving examples and test cases.

Very large eddy simulations of the flow within a F-35's weapon bay with an internal store

Ron Efrati* and Hadar Ben-Gida†
Israeli Air Force, Tel Aviv 6473428, Israel

Modern fifth-generation jet fighters are known to have internal store carriage capabilities (i.e., weapon bays) to improve performance and stealth. When the weapon bay doors are open, an unsteady flow regime is generated, which can adversely affect the safe separation of internal carriage stores as well as the store's electronic systems. Predicting the unsteady flow within realistic weapon bays, such as the one of the F-35, is challenging. High-fidelity tools are required, such as large eddy simulations (LES), which are not well suited for solving complex geometries, such as the weapon bay of a realistic jet fighter, in realistic conditions and within a short period of time. Recently, the Lattice Boltzmann Method (LBM) solver, PowerFLOW, has proven capable of resolving the unsteady flow field within a weapon bay of a benchmark jet fighter with an internal store model. In this study, we further expand the application of the relatively new modeling approach utilized in PowerFLOW, namely the very large eddy simulation (VLES), to resolve the unsteady flow field within the weapon bay of the F-35 jet-fighter with an internal store placed inside. Various parameters are explored in this study for their effect on the flow field and store airloads, such as the mesh resolution, meshing strategy, angle of attack, Mach number and Reynolds number (wind tunnel scaling). The computational results are compared to wind tunnel results, revealing a good agreement in terms of the internal store airloads.

*Aerodynamics Engineer, Aeronautical Engineering Branch, rone5611@gmail.com.
†Aerodynamics Subject Matter Expert, Aeronautical Engineering Branch, bengida@idf.il.

Manuscript 71 submitted to 64th Israel Annual Conference on Aerospace Sciences (IACAS). Received January 9, 2025.

Terrain Ridgelines Detection Based on Streamlines

Baruch E. Karlin, PhD
Research Scientist, Haifa, ISRAEL

I. Problem Statement

Terrain assessment is prevalent in all walks of life–hiking trails to flowfield analysis. Ridgelines are prominent terrain features. Recognition methods were published based on topology, contour curvature, template matching, and image processing. Ridgelines are watersheds–water flows downhill into creeklines (negative ridgelines). Streamlines are geodesics–the shortest local path between two surface points. They are tangent to gradient vectors and perpendicular to contour lines at their intersection points. Goal: bivariate surfaces ridgelines recognition.

II. Methodology

Main steps: Calculate gradient at all grid points; Extract critical points; Derive origins from local contours near each point; Calculate streamline from each origin; Determine ridge/creek type. The procedure was implemented in Matlab® using newly developed functions (see Figures):

- *BEKgrad2Fun*–Reconstruct bivariate function from gradient data.
- *BEKsegCont*–Unpack contour data into segments.
- *BEKonSeg*–Verify point on segment.
- *BEKcumArcInterp*–Nonmonotonic data interpolation (Karlin, IACAS-63, 2024).
- *BEKgetCritPts*–Zero gradient at critical points: Calculate gradient components zero-contours (*contour*); Extract vertices (*BEKsegCont*); Intersect zero-contours (*polyxpoly*); Identify type (extrema/saddle) using Hessian eigenvalues.
- *BEKcontPeaks*–Calculate local contours around extrema; Find positive/negative peaks of radii from extrema point (*findpeaks*); Intersect local circle with peak-radii and with gradient zero-contours near saddle point (*polyxpoly*).
- Set intersections as streamlines' origins.
- Calculate streamlines (*stream2*) for positive/negative gradient field.
- Calculate streamlines vertices heights (*interp2*).
- *BEKisRidge*–vertices higher than their offset neighbors: "ridge", otherwise: "creek".

III. Performance

Testcase terrains: simulated, reconstructed from velocity flowfield, extracted from Digital Terrain Models (DTM). Surface data is assumed error-free for analysis. Hence, assessment deals with how algorithmic assumptions affect the results, including numerical procedures and choice of internal control parameters. For example, contours are piecewise linear approximations with vertices on grid cell edges. This introduces estimation errors to the critical points locations via the intersection of the zero-gradient curves. Streamline endpoint errors may be chaotic. Critical points identification is robust. Good algorithmic performance is demonstrated. Ridgelines originating near saddle points are less sensitive to small variations than those originating near extrema points. Results may vary with grid density. The procedure is a powerful tool for flowfield analysis. Analysis of sensitivity to input data errors is recommended for specific usage. Future work: improved control and special cases.

Figure 1: Gradient Field

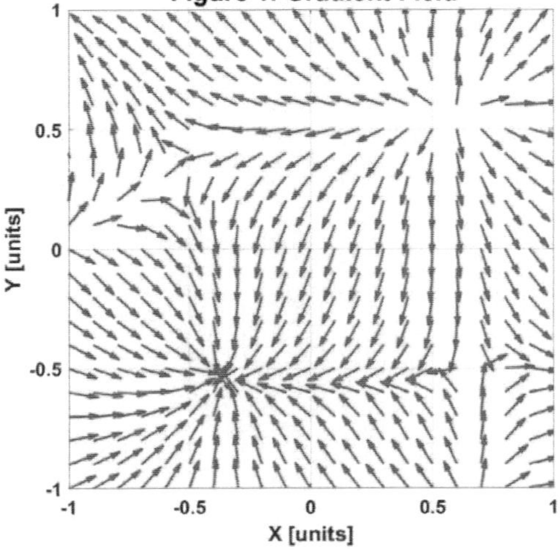


```
Function F = BEKgrad2Fun(X,Y,Fx,Fy,Foo)
%Bivariate function reconstruction from
%gradient data
%X,Y-meshgrid nodes
%hx,hy-grid steps
%[Fx,Fy]=gradient(Z,hx,hy)
%Py(:,1)-initial values for X-integration
%Px(1,:)-initial values for Y-integration
%F-function (assumed continuous)
%Foo-integration constant
%B.E.Karlin, PhD, 15-Dec-2024
[ny,nx] = size(X);
Px(1,:) = cumtrapz(X(1,:),Fx(1,:),2);
Py(:,1) = cumtrapz(Y(:,1),Fy(:,1),1);
Hx = cumtrapz(X(1,:),Fx,2)+repmat(Py,1,nx);
Hy = cumtrapz(Y(:,1),Fy,1)+repmat(Px,ny,1);
F = Foo+(Hy+Hx)/2;
return
```

Figure 2: Terrain with Contours

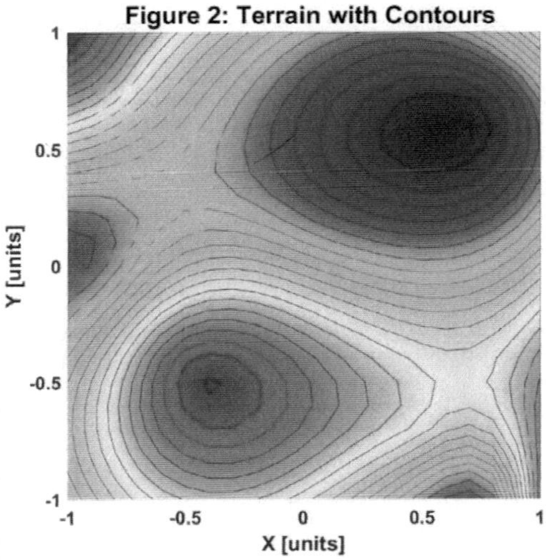


```
function ri = BEKcumArcInterp(r,Nd)
%Nnonmonotonic-data cumulative-arc
%interpolation
%r-mat(dim,:), ri-mat(dim,:), dim=1,2,3
%Nd-subdivisions
%B.E.Karlin, PhD, 25 Dec 2024
s = [0,cumsum(sqrt(sum(diff(r,1,2).^2,1)))];
t = s;
for k=1:Nd %subdivide arc sections
    t = sort([t,(t(1:end-1)+t(2:end))/2]);
end
ri = makima(s,r,t);
return
```

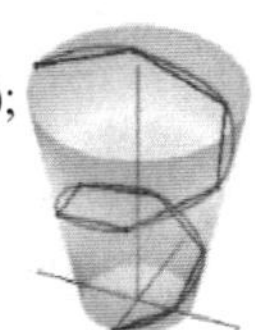

```
function TF = BEKonSeg(xq,yq,xv,yv)
%Verify points on linear segments
%xq,yq-query points
%xv,yv-curve vertices (nonpolygonal)
%TF-verified segments: True/False
%B.E.Karlin, PhD, 23 Dec 2024
if length(xv)>2 %Eexclude end-start segment
    xv = [xv(1:end-1);NaN;xv(end-1:end)];
    yv = [yv(1:end-1);NaN;yv(end-1:end)];
end
[~,TF] = inpolygon(xq,yq,xv,yv);
return
```

Figure 3: Streamlines from Grid Points

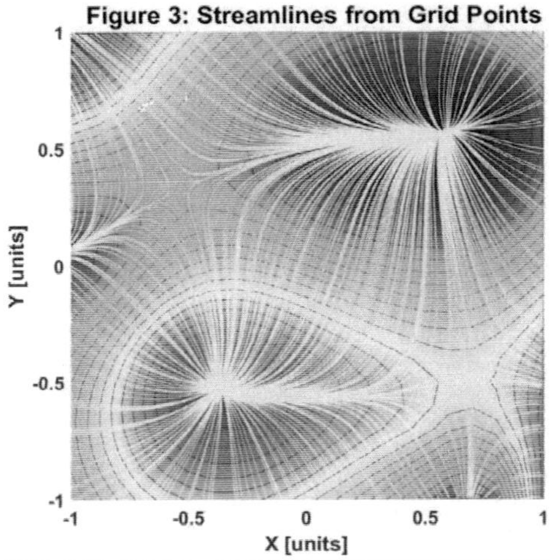

Figure 4: Streamlines from Critical Points

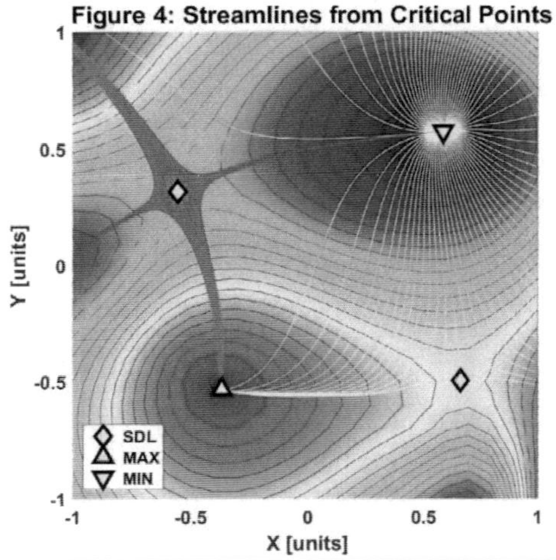

Figure 5: Terrain with Ridgelines

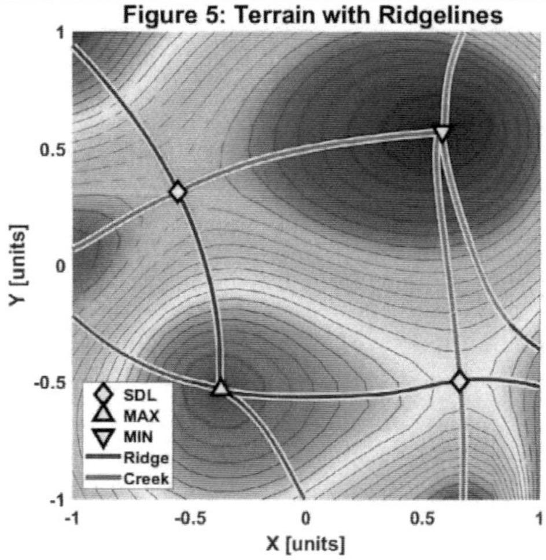

Figure 6: 3D Terrain with Ridgelines

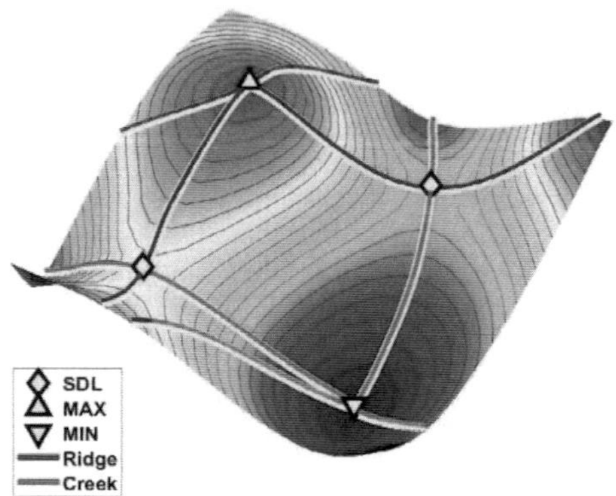

Figure 7: Source-Sink Ridgelines Flowfield with Obstructions

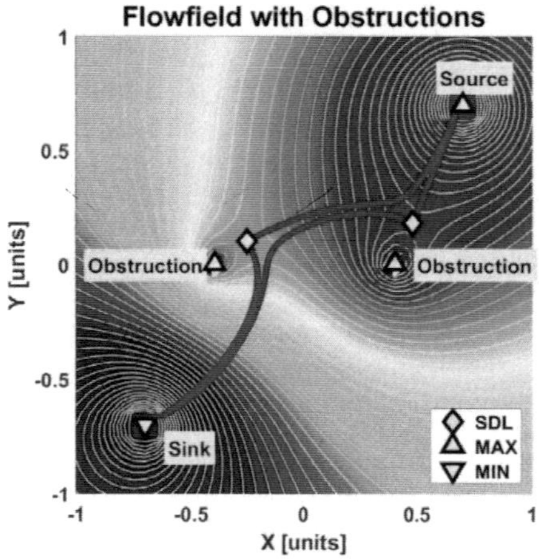

Figure 8: CN vs Mach and Alpha

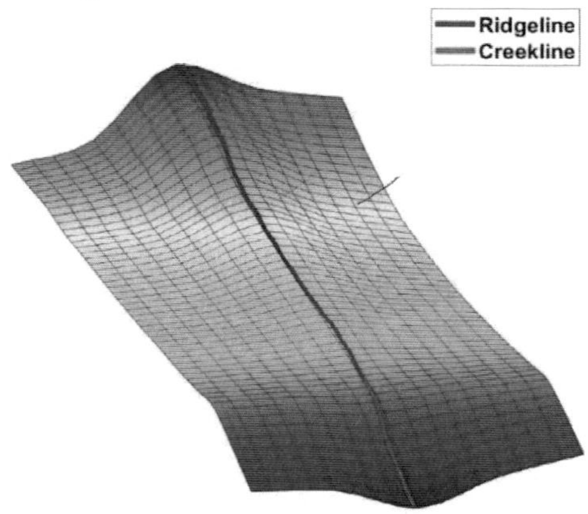

Figure 9: Mount St. Helens Caldera

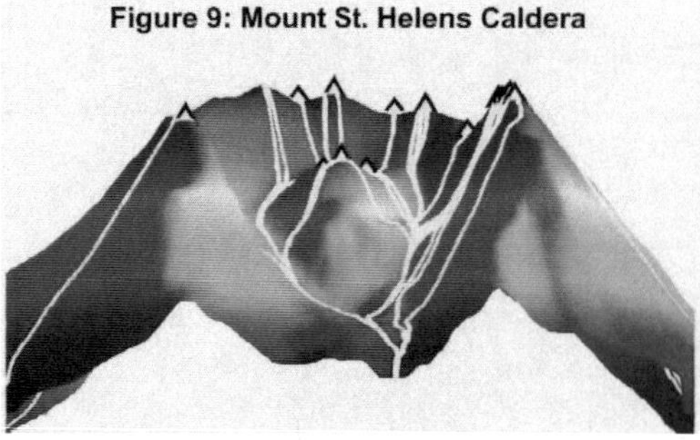

Photo by Lyn Topinka, 1982, Public Domain

Analytical Flow Solutions in Rectangular Microfluidic Channels Based on the Darcy-Brinkman Model

Mario Del Mastro[1] and Alexandros Terzis[2]

***Abstract*— This study examines the application of the Darcy-Brinkman [1] model for deriving analytical solutions to laminar flow in rectangular Hele-Shaw microchannels under no-slip and slip boundary conditions. Two configurations are analyzed: the classical Hele-Shaw cell with impermeable lateral walls (no-slip) and a slip-induced configuration featuring a permeable lateral wall with a structured porous medium, configurations reported in Figure 1. The research highlights the advantages of the Darcy-Brinkman approach in modeling microfluidic systems by providing closed-form solutions that overcome the convergence challenges associated with traditional Fourier series methods.**

For the no-slip configuration, the model effectively captures the boundary layer effects and central plug flow typical of Hele-Shaw systems. Analytical solutions reveal a seamless transition between regions dominated by viscous effects near the walls and uniform flow in the channel's core. Validation with experimental velocity data from micro-PIV [2] demonstrates excellent agreement, particularly for channels with high width-to-height ratios, $w/h > 4$. In this region, the permeability aligns with established formulations for ideal Hele-Shaw cells. The analytical solution for the CHSC configuration is reported in Equation (1).

$$u_x(y, h/2) = \frac{\xi}{\chi^2}\left[1 - \frac{\cosh(\chi(y - w/2))}{\cosh(\chi \cdot w/2)}\right]. \tag{1}$$

The slip-induced configuration introduces an effective slip length L_s [3], which quantifies the interplay between the free-flow and the porous medium at the interface. Analytical and experimental results show that L_s depends on porosity and the geometric properties of the porous structure at the interface between the two domains. Derived models relate L_s to porosity using an inverse error function, enabling accurate predictions of interfacial flow dynamics. Experimental data confirm that higher porosities lead to deeper penetration of free-flow into the porous medium, characterized by increased slip lengths. The analytical solution for the SIC configuration is reported in Equation 2.

$$\langle u_x\rangle(y, h/2) = \frac{\xi}{\chi^2} + \frac{-L_s\chi\xi e^{w\chi} - \xi e^{w\chi} + \xi}{L_s\chi^3 e^{2w\chi} + L_s\chi^3 + \chi^2 e^{2w\chi} - \chi^2} \cdot e^{\chi y} + \frac{-L_s\chi\xi e^{w\chi} - \xi e^{2w\chi} + \xi e^{h\chi}}{L_s\chi^3 e^{2w\chi} + L_s\chi^3 + \chi^2 e^{2w\chi} - \chi^2} \cdot e^{-\chi y}. \tag{2}$$

Comparisons between the Darcy-Brinkman model and other analytical solutions, such as the Spiga-Morini [4] formulation, reveal that the former provides equivalent accuracy while offering computational simplicity. This is particularly evident in high aspect ratio systems, where the Darcy-Brinkman model reliably predicts velocity profiles, boundary layer thickness, and central plug flow behavior.

This research underscores the versatility of the Darcy-Brinkman model for applications in microfluidic systems, including lab-on-a-chip devices and filtration technologies. By addressing challenges in modeling coupled free-flow and porous media systems, it establishes a robust framework for analyzing momentum transfer at interfaces. Future work is proposed to extend the model's applicability to more complex geometries and boundary conditions, further enhancing its utility in engineering and biomedical applications.

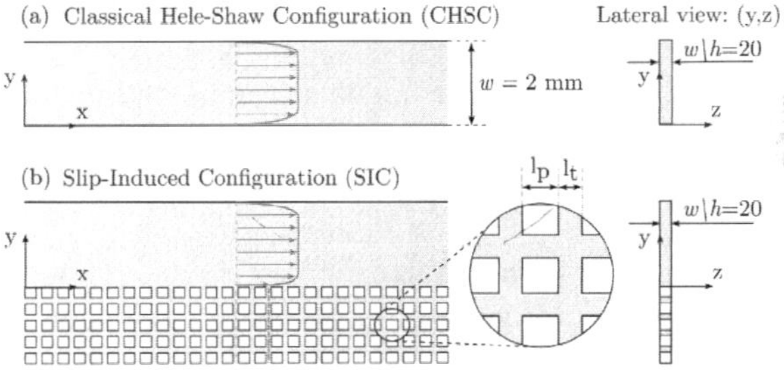

Fig. 1. Schematic of the micromodels with the geometric setup and the coordinate system used for the analysis. Both configurations are characterized by a width-to-height ratio (w/h) of 20 in the (y, z) cross-sectional plane. (a) The Classical Hele-Shaw Cell (CHSC) with no-slip boundary conditions at $y = 0$ and $y = w$. (b) The Slip-Induced Configuration (SIC), which incorporates a porous structure at $y = 0$, designed as a square array of cubic elements with a side length l_p and variable spacing l_t.

References

[1] H. C. Brinkman, "A calculation of the viscous force exerted by a flowing fluid on a dense swarm of particles," Appl. Sci. Res., vol. 1, pp. 27–34, 1949.

[2] J. G. Santiago, S. T. Wereley, C. D. Meinhart, D. J. Beebe, and R. J. Adrian, "A particle image velocimetry system for microfluidics," Exp. Fluids, vol. 25, pp. 316–319, 1998.

[3] C. L. Navier, "Mémoire sur les lois du mouvement des fluides," in Mém. de l'Acad. des Sci. de l'Inst. de France, Paris, 1827, pp. 389–440.

[4] G. M. M. Spiga, "A symmetric solution for velocity profile in laminar flow through rectangular ducts," Int. Commun. Heat Mass Transfer, vol. 21, pp. 469–475, 1994.

*This work was supported by Israel Science Foundation (ISF)

[1] Mario Del Mastro is with Faculty of Aerospace Engineering, Technion – Israel Institute of Technology, Haifa, Israel mario.del@campus.technion.ac.il

[2] Alexandros Terzis is with the Faculty of Aerospace Engineering, Technion – Israel Institute of Technology, Haifa, Israel aterzis@technion.ac.il

Manuscript 42 submitted to 64th Israel Annual Conference on Aerospace Sciences (IACAS). Received January 6, 2025.

Stability Analysis of The Non-Isentropic One-Dimensional Flow in a Scramjet Isolator

S. Haiman* and M. Karp*

In recent years, hypersonic travel is receiving renewed attention. Hypersonic flight has many challenges, especially in the aspects of aerodynamics and propulsion, since they are highly coupled at these speeds. Scramjet engines are a promising propulsion system that enables traveling at such speeds. According to Smart (2010), during the operation of these engines, the disturbances generated by the heat release in the combustor can propagate upstream and affect the performance of the inlet. Therefore, to alleviate this problem, a short duct between the inlet and the combustor is installed, which is commonly known as an isolator. The investigation of disturbances in compressible flow through nozzles is traced back to the work of Tsien (1952). Building on his work, several models have been developed, with a key assumption that the steady base flow is isentropic. Although methods to incorporate non-isentropic phenomena, such as friction, heat transfer, and combustion, have been proposed, their solutions have traditionally relied on asymptotic expansions.

In this study, the instability mechanisms of the non-isentropic flow in an isolator are investigated. The compressible inviscid flow in a scramjet isolator is modeled as a quasi-one-dimensional flow governed by the Euler equations, accounting for effects such as cross-sectional area changes, friction forces, and heat addition by combustion. This model establishes the base flow for analyzing its instability mechanisms. Unlike previous approaches, this work focuses on analyzing the stability of the flow using linear stability analysis rather than asymptotic expansions. Linear stability analysis is a theoretical method for investigating the behavior of infinitesimal disturbances in fluid flows. This approach enables the identification of critical parameters, such as frequency and wavenumber, which can be used to manipulate the flow, and affect the disturbance propagation. Our study emphasizes the temporal stability of this flow, in which the frequencies of disturbances are of primary interest. A modal analysis is conducted, revealing the dominant perturbations, which are classified into acoustic (isentropic), vorticity and entropy waves.

References

SMART, MICHAEL K. 2010 Scramjet isolators. *AVT-185 AVT/VKI Lecture Series, von Kármán Inst. RTO-EN-AVT-185-AC/323 (AVT-185) TP/375, Rhode St. Genèse, Belgium* **0**, 1–10.

TSIEN, HSUE-SHEN 1952 The transfer functions of rocket nozzles. *Journal of the American Rocket Society* **22** (3), 139–143.

*Faculty of Aerospace Engineering, Technion – Israel Institute of Technology, Haifa 32000, Israel

Manuscript 44 submitted to 64th Israel Annual Conference on Aerospace Sciences (IACAS). Received January 6, 2025.

On linear stability of a viscous shock layer

Vassilis Theofilis[1*]

The internal structure of a steady (laminar) fully-resolved viscous shock layer of an ideal gas in the continuum regime has been described by Gilbarg and Paolucci [1953]. Somewhat surprisingly, the only work known to-date to have addressed linear stability of this flow has been that of Duck and Balakumar [1992]. These authors, henceforth referred to as DB92, derived two-dimensional temporal linear stability equations (describing shock corrugation on the shock plane) and reported that no discrete exponentially-growing two-dimensonal eigenmodes could be identified. DB92 also commenced analysis of the corresponding initial-value problem and provided an analytic demonstration of the long-time structure of perturbations, as products of arbitrary functions of time and the exponentially decaying (in time) factor of modal analysis. However, complete solutions of the initial-value problem could not be found analytically, due to contamination of the spectrum by spurious numerical modes. The present work has revisited the DB92 analysis, also in a temporal context, extended the linearized disturbance equations in the third spatial dimension (into the plane along which the base flow is homogeneous) and re-introduced the Reynolds number (which was absorbed into the governing equations by DB92) explicitly into the governing equations. The three-dimensional eigenvalue problem has been solved for a sufficiently large number of discrete values of the pertinent parameters (combinations of transverse and lateral wavenumbers at given values of the Mach and Reynolds numbers) for the structure of the spectrum to be established and the influence of the wavenumber parameters to be understood. Systematic scan of Mach and Reynolds numbers has followed, to address the effect of these parameters on weak shock layer stability. Finally, the corresponding initial-value problem was solved and revealed the early-time response of the shock layer to small-amplitude perturbations.

References

P. W. Duck and P. Balakumar. On the stability of normal shock waves. In M. Y. H. et al., editor, *Instability, Transition and Turbulence*, pages 253–265. Springer Verlag, NY, 1992.

D. Gilbarg and D. Paolucci. The structure of shock waves in the continuum theory of fluids. *J. Rat. Mech. and Anal.*, 2:617–642, 1953.

*Work supported by ONR Grant N00014-23-1-2839 *Kinetic Treatment of Sources and Mechanisms that Drive Unsteady, Shock-dominated Flow Instability* with Dr. Eric Marineau as Program Officer

[†1]Faculty of Aerospace Engineering, Technion – Israel Institute of Technology vassilis@technion.ac.il

Manuscript 45 submitted to 64th Israel Annual Conference on Aerospace Sciences (IACAS). Received January 6, 2025.

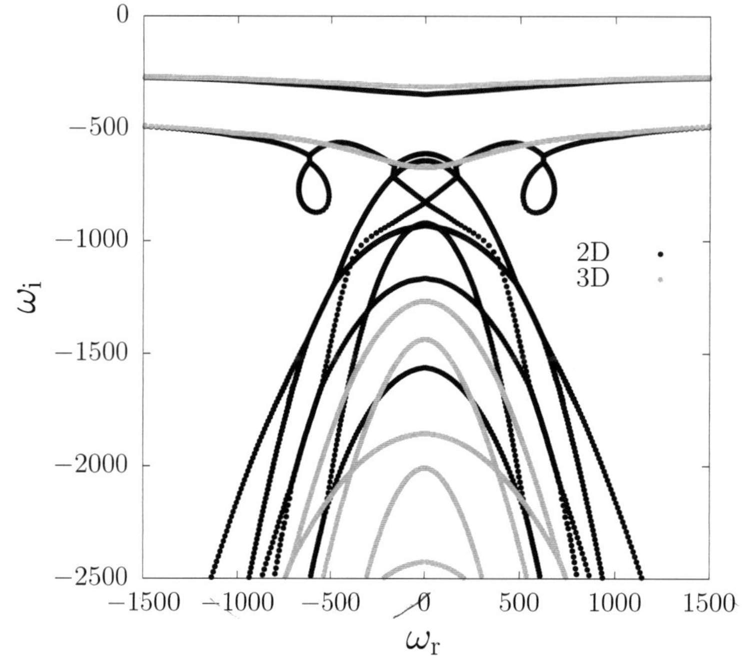

Figure 1. Two- vs. three-dimensional eigenspectrum at $Ma = 1.5$

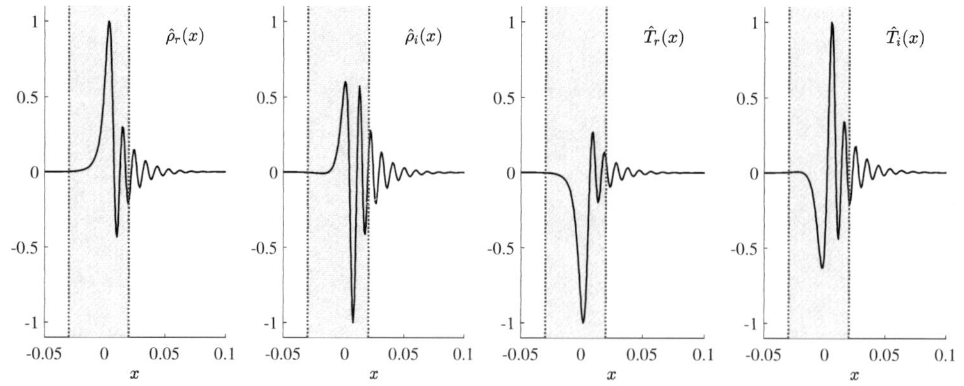

Figure 2. Amplitude functions of a damped mode at $Ma = 1.5, Re = 10^3, \beta = 960, \delta = 0$ showing the real (*left column*) and imaginary (*right column*) parts of the density, $\hat{\rho}$ and temperature, $\hat{T}$, components of the disturbance eigenvector.

Manuscript 45 submitted to 64th Israel Annual Conference on
Aerospace Sciences (IACAS). Received January 6, 2025.

Stability analysis of forced and unforced flow fields around cylinders in crossflow

Doron Schwartz* and Michael Karp*

The cylinder in cross-flow is a canonical problem for investigation of flow separation. Injection of a wall-jet at the surface of the cylinder is one of the possible methods to control the flow. The jet adds momentum to the boundary layer, thus delaying the separation. This not only alleviates the pressure drag on the cylinder, but also generates lift, since the separation pattern is not symmetric. One of the most celebrated practical applications is the NOTAR helicopter (Hanvey, 1982), where the tail rotor is replaced by tangential slot blowing and a directional jet.

The wall-jet profile is characterized by a local maximum of the tangential velocity near the surface. The jet tends to follow the surface of the cylinder, where the pressure is lower than its surrounding value, in accordance with the Coandă effect (Coandă, 1936). Much of the knowledge gained on this problem was obtained by examining two-dimensional slot blowing. Recent findings demonstrate the potential to utilize an array of three-dimensional fluidic oscillators rather than a two-dimensional slot.

For flows over curved surfaces, a basic mechanism of instability arises due to the centrifugal force, known as the Görtler instability (Saric, 1994). The instability manifests itself in the form of Görtler vortices, which are steady, streamwise-oriented, counter-rotating vortices. Wall-jets on convex surfaces are prone to Görtler instabilities, since the velocity in the outer region decreases away from the wall. The Görtler instability has the potential to be utilized for flow control purposes, by using an array of wall-jets on the surface of a cylinder. The wall-jets' spanwise spacing needs to match the theoretically-predicted spanwise wavenumber.

The current study aims to conduct a theoretical analysis of measured velocity fields, by means of Particle Image Velocimetry, obtained in the Fluid Mechanics Research Laboratory led by Prof. Ari Glezer at the Georgia Institute of Technology. The theoretical analysis includes the local Linear Stability Theory (LST) and Parabolized Stability Equations (PSE), where the latter takes into account the streamwise variation of the flow along the surface of the cylinder. The analysis reveals the dominant frequencies and spanwise wavenumbers of the perturbations, which is essential in order to inform the flow control and guide future experiments.

References

COANDĂ, H. 1936 Device for deflecting a stream of elastic fluid projected into an elastic fluid. US Patent 2,052,869.

HANVEY, S.A. 1982 No tail rotor (circulation control tail boom). In *26th Annual Symposium of The Society of Experimental Test Pilots*, p. 1.

SARIC, W. S. 1994 Gortler vortices. *Annual Review of Fluid Mechanics* **26** (1), 379–409.

*Faculty of Aerospace Engineering, Technion – Israel Institute of Technology, Haifa 32000, Israel

Manuscript 46 submitted to 64th Israel Annual Conference on Aerospace Sciences (IACAS). Received January 6, 2025.

Particle Clustering and Turbulence Modulation in Eulerian Particle-Laden Flow Simulations

Ajay Dhankarghare* and Yuval Dagan†
Technion - Israel Institute of Technology, Haifa, 3200003, Israel

Turbulent flows laden with dispersed particles exhibit different dynamic behaviors compared to the corresponding clean flows. Adding particles to the flow leads to various phenomena like inter-particle collisions, particle-flow interactions, wall-particle collisions, particle clustering and migration, and flow turbulence modulation. In a two-way coupled system, while particles can significantly modulate the flow turbulence, flow turbulent structures can affect the particles' behavior like clustering. However, understanding this interplay between particle clustering and turbulence remains elusive. Additionally, computational studies of flows with large numbers of particles are challenging due to their high computational cost. In the present study, a two-way coupled, GPU-accelerated solver to perform Direct Numerical Simulation of particle-laden compressible flows has been developed. The fluid phase is solved using the SLAU2 upwind scheme that has been modified to handle both high and low Mach number flows. The solver's ability to simulate the clean turbulent flow at a low Mach number of 0.13 has been validated against previous DNS results[a]. The particle phase is solved using another upwind scheme[b]. In contrast to prior research that employed a Lagrangian approach for particle flow simulations, this study[c] integrates an Eulerian approach based on the quadrature-based moment method, thus facilitating the analysis of flows laden with an extremely large number of particles.

The solver is applied to simulate dispersed particle laden turbulent flow ($Re = 5600$) in a rectangular channel with particle-wall elastic reflection. Three friction Stokes numbers ($St^+ = 1, 6$ and 30) are studied at various particle mass loadings (ϕ_m). We find that the addition of particles to the clean flow affects the mean flow rate and the flow turbulence. In the presence of turbulence, the particles migrate to the near-wall region. This near-wall migration of particles is largely influenced by St^+ and ϕ_m. Furthermore, the particles tend to preferentially cluster in the fluid velocity streaks. In doing so, they can affect the velocity streaks and their distribution, which can finally modulate the turbulence. We further study the interaction of particles with turbulent eddies. Due to inertia, the particles tend to migrate towards the periphery of these eddies. This migration of the particles may depend on St^+ which represents the particle inertia. The interaction of particles with the eddies can also affect the length scale, distribution, and dissipation of these eddies. By simulating the Eulerian particle field in a turbulent channel flow, we explore the interplay between particles and fluid velocity streaks; and between particles and turbulent eddies, leading to the modulation of turbulence.

---

*Ph.D. Scholar, Faculty of Aerospace Engineering, ajayd@campus.technion.ac.il
†Assistant Professor, Faculty of Aerospace Engineering, yuvalda@technion.ac.il
[a]Kim et al., *Journal of Fluid Mechanics* **177** (1987).
[b]Desjardins et al., *Journal of Computational Physics* **227** (2008).
[c]Dhankarghare and Dagan, *https://arxiv.org/abs/2409.07988*

Manuscript 82 submitted to 64th Israel Annual Conference on Aerospace Sciences (IACAS). Received January 13, 2025.

Investigation of oil flow topology, pressure distribution, and drag in separated flow over low-Reynolds-number airfoil

Roi Baruch* and Igal Gluzman†
Technion – Israel Institute of Technology, Haifa 3200003, Israel

I. Abstract

The flow behavior over an airfoil at low Reynolds numbers (at typical chord Reynolds numbers up to 500,000) is fundamentally more complicated than at high Reynolds numbers. At these low Reynolds numbers, the laminar boundary layer may exhibit a non-abrupt transition to turbulence during flow separation. The initial laminar boundary layer may detach from the airfoil surface, leading to an unstable separation bubble that leads to degradation in performance at low Reynolds numbers and increases the drag. This study is the first part of our primary goal to integrate actuators into the existing design of low Reynold airfoils to obtain high-performance, maneuverable airfoils via active flow control. Here, we focused on establishing baseline data over existing airfoil design (NACA 63-412) for assessing different active flow control strategies on transition control and reducing the size and extent of the separation bubble. We conducted wind tunnel experiments at various angles of attack (up to 14 deg), measuring the surface pressure distribution around the airfoil and at its wake to estimate the drag force on an airfoil. In addition, detailed surface oil flow visualizations and oil-film interferometry (OFI) are employed for accurate skin friction measurements and quantitative information on the surface flow topology via utilizing our simplified photogrammetry procedure [1,2] for projecting oil flow visualizations from 2D images over a 3D surface. In this talk, we will discuss the relations between critical points of surface topology, estimated drag, and pressure distribution. In particular, relations have been established between the first and second spatial derivatives of the streamwise static surface pressure profile to detachment and reattachment locations.

References

1. Gluzman, I., Gray, P., Mejia, K., Corke, T. C., and Thomas, F. O., "A simplified photogrammetry procedure in oil-film interferometry for accurate skin-friction measurement over arbitrary geometries," *Experiments in Fluids*, Vol. 63, No. 7, 2022, pp. 118.

2. Gluzman, I., Gray, P., Corke, T. C., and Thomas, F. O., "A simplified photogrammetry procedure for quantitative on-and-off surface flow topology studies over 3D surfaces: application to a Gaussian bump geometry," *Experiments in Fluids*, Vol. Under review, 2024.

*Dept. of Aerospace Engineering. roi.baruch@campus.technion.ac.il
†Asst. Prof., Dept. of Aerospace Engineering. igal.gluzman@technion.ac.il

Manuscript 67 submitted to 64th Israel Annual Conference on Aerospace Sciences (IACAS). Received January 9, 2025.

Pressure Effect on the Burning Rate of Solid Propellants with Expandable Graphite Additives

Noa Edri and Alon Gany

Solid propellant motors have been widely used because of their simplicity, ease of operation, and readiness for firing. Solid propellants exhibit characteristic burning rate related to the propellant formulation and pressure according to the common correlation

$$\dot{r} = aP^n$$

where a and n depend on the propellant formulation, while a is also slightly affected by the initial temperature. For a certain geometry, an increase of the characteristic burning rate of a given propellant, can be done by specific catalysts.

Recently, expandable graphite (EG) additives have been investigated in the Fine Rocket Propulsion Center of the Technion. The effect of EG on the combustion in a hybrid motor was studied by *Elanjickal and Gany [1]*, revealing up to a 100% increase in the fuel regression rate. A close view of the EG behavior on the surface of fuel slabs subjected to flame has been presented by *Muller and Gany [2-3]* and by *Muller et al [4]*. Expandable graphite is a form of intercalated graphite where a small fraction of guest molecules is introduced among the graphite crystal layers. The main property of EG is its expansion, forming worm-like strings longer by an order of magnitude than the original particles, when heated to temperatures higher than about 200^0C. It was hypothesized that when approaching the hot surface layer during combustion, the EG strings would penetrate from the surface to the hot surrounding environment and transfer heat by conduction into the fuel bulk, thereby increasing the burning rate. Initial investigation on the effect of EG additives on the burning rate of solid propellants took place by *Muller and Gany [5]*, indicating the possible application of EG for enhancing the burning rate of solid propellants. Conducting experiments of propellants composed of ammonium perchlorate (AP) and hydroxyl-terminated polybutadiene (HTPB) and others composed of AP and polyester, at atmospheric pressure, revealed that the former ones were largely affected by the addition of up to 5% EG, whereas the latter were practically not affected. However, solid rocket motors operate at much higher pressures, and no publication in the world has ever presented research on the effect of EG additives on the burning rate of solid propellants in such conditions.

Our research studies experimentally the effects of EG additives on the burning rate of propellants at high pressures. The test setup includes a controlled high pressure, widowed cylindrical cell accommodating a solid propellant strand (Fig. 1). The burning rate of the strand is determined by measuring the time for the flame propagation between fuse wires that burn and disconnect when approached by the flame. An electronic setup connected to a computer with graphical user interface (GUI) send an ignition command accomplished by an electric ignition wire. This enables obtaining the real-time pressure and burning rate data. Propellant strands of different compositions (mainly AP-HTPB and AP-polyester) are used. Experiments are conducted at different pressures up to about 70 atm (7 MPa). At each pressure level, series of burning rate experiment with different EG types and different EG fractions (from zero to

Manuscript 21 submitted to 64th Israel Annual Conference on Aerospace Sciences (IACAS). Received December 22, 2024.

about 5%) take place. The propellant strands are prepared in a designated mold with different types and quantities. Typical strand will be of rectangular 7 mm x 7 mm cross-section and 40 mm long.

Figure 1: The high-pressure windowed test cell.

The current results reveal an improvement of approximately 10% in the burning rate for propellants containing 3% EG Type ES 100 C10 of Graphit Kropfmühl GmbH, Germany, nominal particle size 150 μ, compared to those without EG, in both polyester-based propellants (Fig. 2a) and HTPB-based propellants (Fig. 2b). Contrary to previous research, this study did achieve an improvement in the polyester-based propellant for a 3% EG additive, as highlighted. However, AP-polyester propellants with 1% EG not only failed to improve the burning rate but also exhibited slower rate at higher pressures. In the future, we intend to conduct additional experiments, investigating other proportions of EG and exploring different particle types.

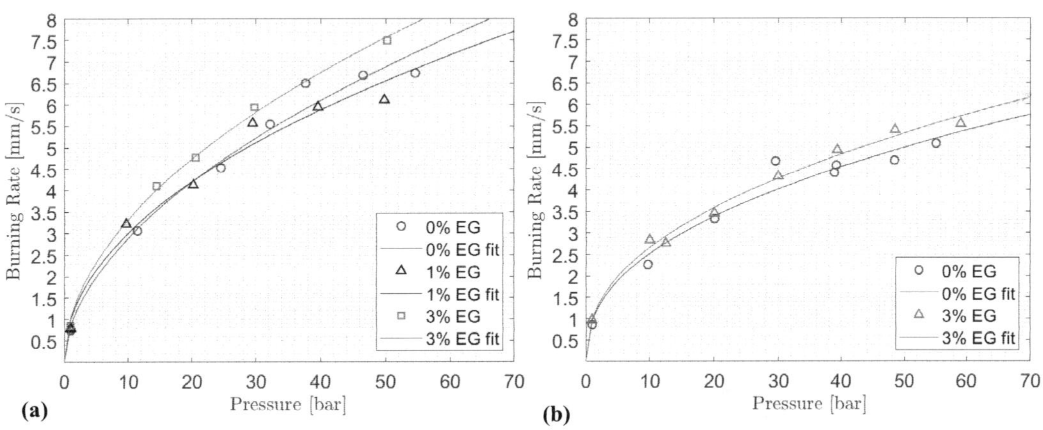

Figure 2: Burning rate as function of pressure for different percentages of EG Type ES 100 C10 . (a) Polyester and AP based propellants (b) HTPB and AP based propellants.

Acknowledgment

This research was supported by PMRI – Peter Munk Research Institute, Technion.

References

1. Elanjickal, S. and Gany, A., "Enhancement of the Fuel Regression Rate in Hybrid Propulsion by Expandable Graphite Additive", a Special Journal Issue for Bill

Sirignano's 60th Anniversary of Academic Research, Combustion Science and Technology, Vol. 192, No. 7, 2020, pp. 1253-1273. doi:10.1080/00102202.2020.1748017.

2. Muller, T.G. and Gany, A., "Burning Phenomena of a Polymeric Fuel Containing Expandable Graphite", *Propellants, Explosive, Pyrotechnics*, Vol 45, 2020, pp. 1764-1768. DOI: 10.1002/prep.202000138.
3. Muller, G.T. and Gany, A., "Expandable Graphite Effect on Solid Fuel and Propellant Combustion", *FirePhysChem,* Vol. 2, 2022, pp. 72-75. https://doi.org/10.1016/j.fpc.2021.12.001
4. Muller, G.T., Zamir, I., and Gany, A., "Thermal Characteristics of Expandable Graphite as a Burning Rate Enhancer in Hybrid Propulsion", *Propellants, Explosives, Pyrotechnics,* Vol. 48, e202300074, Open Access, July 2023. https://doi.org/10.1002/prep.202300074
5. Muller, G.T. and Gany, A., "Increasing Burning Rate and Motor Thrust by Expandable Graphite Additives", *Journal of Propulsion and Power,* Vol. 39, No. 5, 2023, pp. 709-717. https://doi.org/10.2514/1.B39126

Investigating a Gun-Launched Solid Fuel Ramjet Projectile

Itamar Levitan, Alon Gany

Technion - Israel Institute of Technology, Haifa, Israel

Conventional means for range extension in Gun-Launched projectiles, such as rocket assisted projectiles (RAP), or Base Bleed provide limited improvement. For this purpose, the Solid Fuel Ramjet (SFRJ) seems like the natural option, enabling Thrust-Equal-Drag flight implying a ballistic (pseudovacuum) trajectory, and to some extent, can even balance out wind deviations (Gany [1]).

In this research we are trying to achieve Thrust-Equal-Drag for a 76 mm ramjet projectile, which flies at Mach 2.7 at sea-level, has a 170 mm length, 60 mm internal diameter hollow cylinder fuel. The literature indicates some flameholding constrains, implying in our case a rearward facing step of 8.8 mm height at the combustor entrance (Netzer and Gany [2]). The calculated aerodynamic drag is 721 N, requiring thrust of the same order.

For static firing we're using a direct-connected ramjet combustor configuration, receiving hot air via a vitiated air-heater (Fig. 1).

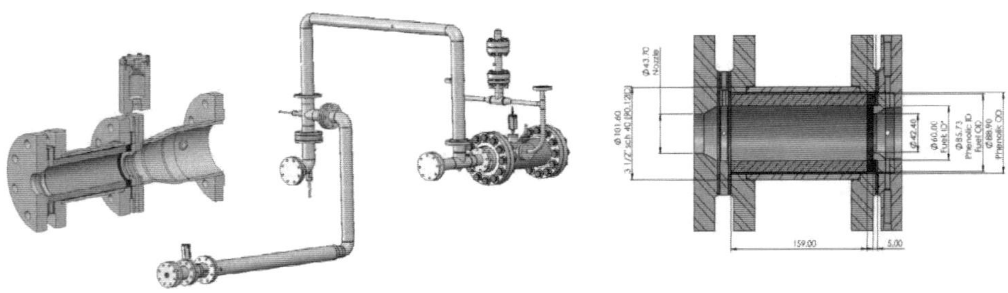

Fig 1. The experimental System and Engine cross section

A series of static firing experiments has been conducted with the baseline fuel, Polyester, varying the airflow rate from 0.5 to 1.45 kg/s. The results revealed the fuel regression rate correlation, as well as the fuel-to-air ratio f and thrust.

Thermochemical calculation helped predicting the required fuel-to-air ratios for thrust-equal-drag conditions[2].

The experimental regression rate results with polyester fuel, showed thrust lower by about 50% than needed, because of low regression rate and low fuel-to-air ratio. The required fuel-to-air ratios at different airflow rates are shown in Fig. 2.

Manuscript 27 submitted to 64th Israel Annual Conference on Aerospace Sciences (IACAS). Received December 24, 2024.

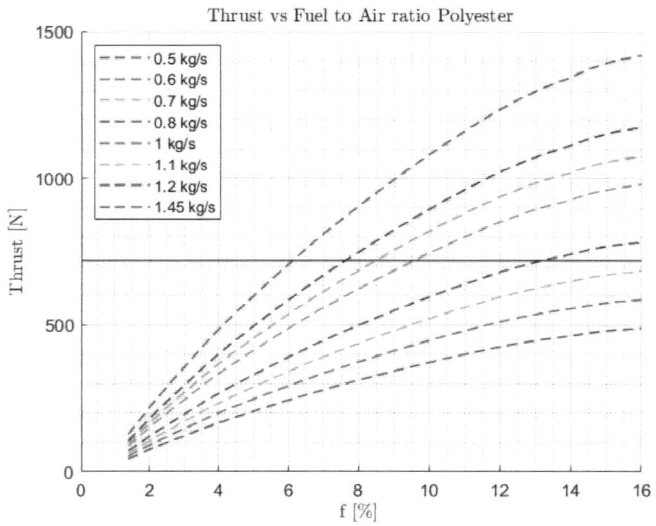

Fig 2. Thrust vs Fuel-To-Air ratio for different airflow rates

Experiments from area of hybrid combustion, indicate that Expandable Graphite (EG) additives could increase the regression rate (Elanjickal and Gany [3], Muller and Gany [4]). With addition of 5% EG, regression rate was increased by 50%, indicating that thrust equal drag conditions might be achieved for an airflow rate of 1.45 kg/s. HTPB fuel was used as well, and in terms of regression rate demonstrated similar performance to Polyester with 5% EG.

Acknowledgement

We acknowledge the support received through the ATS Fund for Applied Security Science and Technology Research.

Tomer's support of the research is highly acknowledged as well.

References

1. Gany, A., "Analysis of Gun-Launched, Solid Fuel Ramjet Projectiles", *International Journal of Energetic Materials and Chemical Propulsion*, Vol. 1. Issue 1-6, 1991, pp. 289-309.
2. Netzer, A. and Gany, A., "Burning and Flameholding Characteristics of a Miniature Solid Fuel Ramjet Combustor", J. of Propulsion and Power, Vol. 7, No. 3, May-June 1991, pp. 357-363.
3. Elanjickal, S. and Gany, A., "Enhancement of the Fuel Regression Rate in Hybrid Propulsion by Expandable Graphite Additive", a Special Journal Issue for Bill Sirignano's 60th Anniversary of Academic Research, Combustion Science and Technology, Vol. 192, No. 7, 2020, pp. 1253-1273. doi:10.1080/00102202.2020.1748017.
4. Muller, G.T. and Gany, A., "Increasing Burning Rate and Motor Thrust by Expandable Graphite Additives", Journal of Propulsion and Power, Vol. 39, No. 5, 2023, pp. 709-717. https://doi.org/10.2514/1.B39126

Enhancing Water-Breathing Ramjet Performance with Boron Additive

Sagi Dinisman and Alon Gany

Technion—Israel Institute of Technology, Haifa 3200003, Israel

A water-breathing ramjet is a rocket-based motor that utilizes the ambient water to enhance the thrust and energetic performance of a high-speed underwater vehicle. At elevated velocities (100-150 m/s), the resulting high ram pressure (50-110 atm) enables the ingestion of water into the motor. When mixed with the hot combustion products of the propellant, the water boils, generating steam that increases the overall mass flow rate and can serve as an oxidizer for hydro-reactive metals.

The highly energetic boron-water reaction presents significant potential for employing boron as an additive in water-breathing ramjets. The features of water-breathing ramjets are particularly suited to harness boron's energy: the high temperature of the propellant favors boron ignition, while the lower temperature section (mixing chamber) allows for effective heat recovery due to condensation of the boron combustion products.

This study presents an extensive series of over 100 static firing tests, evaluating the performance of a water-ramjet utilizing boron additive. The results, shown in Fig.1, demonstrate a maximum actual standard specific impulse of 510 s with a combustion efficiency of 90%. Compared to commonly employed metals such as magnesium and aluminum, boron shows a 10% improvement in performance across broad ranges of water-to-propellant mass ratio, thereby expanding the operational range of the motor. Typical experiment results are shown in Fig. 2, detailing (a) the water mass flow rate and (b) thrust. The experiment is divided into two main stages. The first stage, highlighted in red, represents the hybrid rocket operation without water entry. The second stage, highlighted in blue, represents the water-breathing ramjet mode. The results indicate a substantial increase in thrust (by a factor of 2.5), demonstrating that boron can efficiently react with water and produce high thrust.

Manuscript 19 submitted to 64th Israel Annual Conference on Aerospace Sciences (IACAS). Received December 22, 2024.

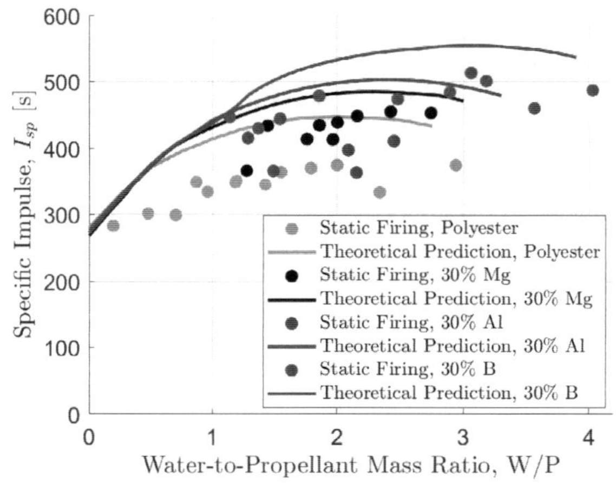

Fig. 1 Standard specific impulse as a function of the water-to-propellant ratio for plain polyester and polyester with 30% mass fraction of boron, magnesium and aluminum

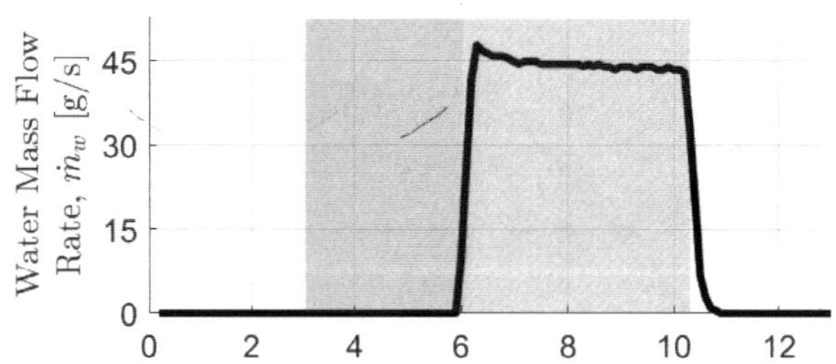

(a)

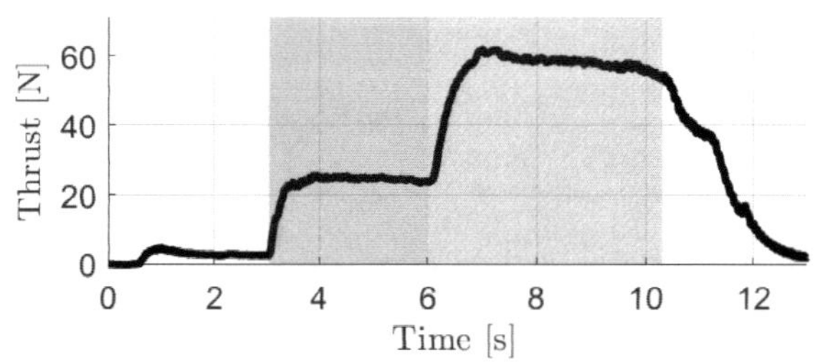

(b)

Fig. 2 Typical results of a single static firing experiment for a water-breathing ramjet employing boron as an additive: (a) water mass flow rate and (b) thrust.

Manuscript 19 submitted to 64th Israel Annual Conference on Aerospace Sciences (IACAS). Received December 22, 2024.

Experimental Study of Mode Transition in a Cavity-based Dual-mode Scramjet Combustor

Dual-mode scramjets integrate the advantageous capabilities of both ramjets and scramjets within a single fixed geometry engine. This combination may allow vehicle travel between Mach 3-4 to Mach 7 and above. Operation using a fixed combustor geometry between Mach 3 to 7 reduces significantly the required weight of the solid propellant booster for reaching hypersonic flight. Moreover, it paves the way for a combined cycle that uses a turbojet or turbo-ramjet engine for acceleration to Mach number of 3 at which the dual-mode scramjet can take over. Transition between different modes of dual-mode scramjet engine is achieved by controlling the heat release. The axial distribution of heat release is especially important in dual-mode scramjet combustors since it determines the conditions under which isolator unstart and ram to scram transition occurs. For ramjet mode operation this distribution affects the location of the thermal throat, and thus the entire distribution of Mach number, pressure, and other flow quantities in the combustor. Due to its effect on overall engine performance and operability limits, it is important to be able to predict the heat release distribution over the range of operating conditions.

This study aims to understand the governing parameters and dynamic combustion behavior during the scramjet-ramjet transition in a dual-mode scramjet with a cavity flame holder. The study is conducted on a water-cooled dual-mode scramjet combustor, enabling relatively long experiments to achieve the transition at a quasi-steady state when the thermal balance between the reacting flow and combustor walls is achieved. The combustor also incorporates a long isolator that can sustain the whole length of the shock train and enable studying the isolator-combustor interaction during the combustion mode transition. This transition process is experimentally investigated by measuring the static pressure along the walls and CH* chemiluminescence imaging. The fuel used in this study is a mixture of ethylene and methane at different mixture ratios. The main emphasis is on the flame stabilization dynamics before, during, and after the mode transition. The study provides insight into the effect of equivalence ratio and fuel composition on the flame stabilization modes and operation mode transition. It is shown that the transition in flame stabilization mode due to a change in equivalence ratio or fuel composition (cavity shear layer vs. jet wake) is accompanied by operation mode transition (scramjet mode vs. ramjet mode).

Hydrogen Propulsion: The Fundamental Equation for H_2 in Its Liquid, Saturated, and Gaseous States

Shir Levi[1] and Eran Sher[2]
[1]Department of Mechanical Engineering, Technion, Haifa
[2]Department of Aerospace Engineering, Technion, Haifa

Abstract

Hydrogen's potential as a high-energy fuel has spurred significant interest in its application in aviation propulsion over the past decades. Understanding combustion characteristics, whether hydrogen is used alone or in combination with other fuels, is critical. Key questions often concern how variations in fuel injection and combustion conditions impact the generated heat output and combustion product composition. Our study focuses on the thermodynamic properties of hydrogen under different conditions.

Accurate values for properties such as specific internal energy, enthalpy, and entropy are essential. While tabulated data can be useful, a fundamental equation offering a comprehensive range of thermodynamic properties presents distinct advantages, especially for properties like those of liquid hydrogen at negative absolute pressures not covered by typical data sources.

The primary challenge is developing this fundamental equation. We have based our approach on the Redlich-Kwong equation of state, supplemented by an additional equation derived from entropy principles, completing the model with a constant third equation. This newly developed fundamental equation has been compared with values from the scientific literature and simpler model simulations, showing strong correlation with published results, particularly within 0.5 to 2.5 times the critical pressure and 2.5 to 10 times the critical temperature, as per the NIST database.

Our results demonstrate the superiority of this development over equations based on the ideal gas law and Van der Waals model, marking a significant advancement in the accurate modeling of hydrogen's thermodynamic behaviors across various states.

Manuscript 74 submitted to 64th Israel Annual Conference on Aerospace Sciences (IACAS). Received January 10, 2025.

Computer Vision Approach for Analysis of Numerical and Experimental Detonation Cellular Structure Images

Daniel Jalontzki[1], Alon Zussman[1], Sumedh Pendurkar[2], Guni Sharon[2], Yoram Kozak[1]

[1] *School of Mechanical Engineering, Tel Aviv University, Tel Aviv, Israel*

[2] *Department of Computer Science & Engineering, Texas A&M University, College Station, TX, USA*

Combustion via gas-phase or liquid spray detonation waves can potentially be used to develop innovative and highly efficient aerospace propulsion systems [1]. On the other hand, detonations also present significant safety hazards [2]. Thus, accurate numerical models to replicate and predict real-world detonation behavior are essential.

While multidimensional numerical simulations of gas-phase and liquid spray detonations are quite common today, validating their accuracy remains an open challenge. A key step for validating the accuracy of numerical simulations is correctly measuring the detonation cell size obtained from soot foils. However, manual measurements are prone to errors exceeding 50%, and automated approaches, while promising, still can be further improved, see, for instance, [3], [4], [5]. Consequently, precise measurements of the detonation cell size remain very challenging.

Addressing these gaps, we introduce a fully automated, computer-vision-based method for measuring and analyzing detonation cell sizes. The proposed framework comprises four key steps: (1) image preprocessing, (2) contour detection, (3) parameter optimization, and (4) statistical analysis. Our approach achieves relative errors of less than 10% for regular and 17% for irregular numerical soot foils. Validation against experimental detonation cell images further demonstrates our method ability to perform detailed statistical analyses across various experimental channel sections.

The presented framework aims to standardize detonation cell measurements by offering a quick and objective soot foil analysis. Future work will further explore the applicability of this approach across a broader range of both numerical and experimental soot foil images.

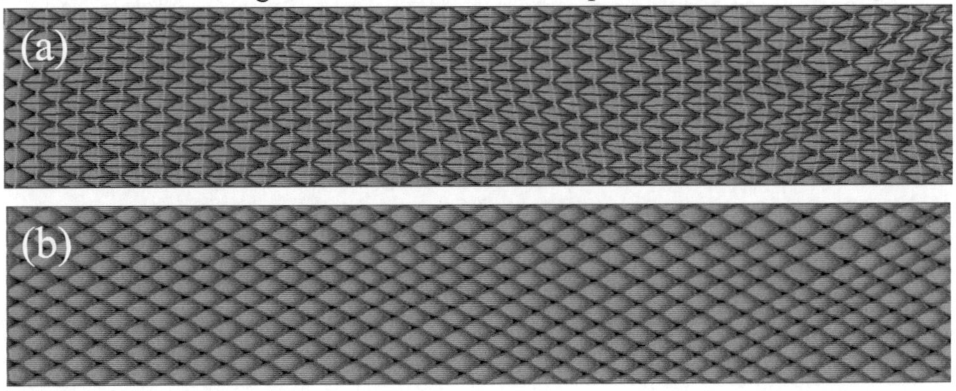

Figure 1: (a) Computer vision approach cell width and length measurements, and (b) contour detection (red lines) of a nearly-regular numerical soot foil image.

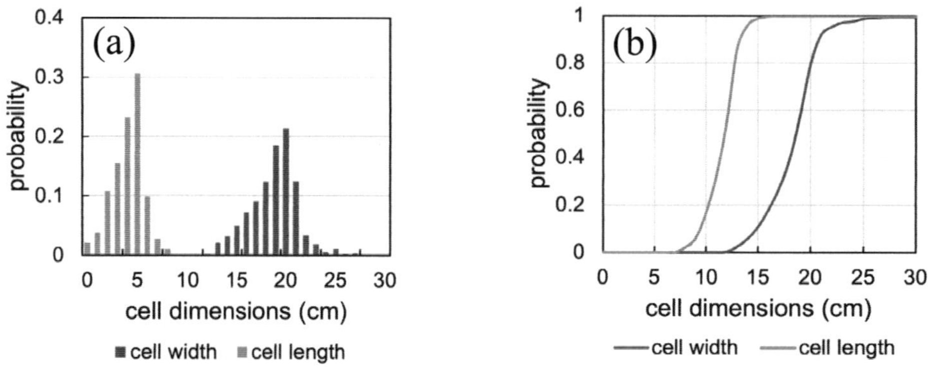

Figure 2: Statistical analysis for a nearly-regular numerical soot foil image, as shown in Fig. 1: (a) Histograms, and (b) CDF.

References

[1] P. Wolański, " Detonative propulsion ," *Proceedings of the Combustion Institute*, vol. 34, no. 1, p. 125-158, 2013, doi: 10.1016/j.proci.2012.10.005.

[2] S. Eidelman and A. Burcat, "Evolution of a detonation wave in a cloud of fuel droplets. I - Influence of igniting explosion," *AIAA Journal*, vol. 18, no. 9, pp. 1103–1109, 1980, doi: 10.2514/3.7711.

[3] G. J. Sharpe and M. I. Radulescu, "Statistical analysis of cellular detonation dynamics from numerical simulations: one-step chemistry," *Combustion Theory and Modelling*, vol. 15, no. 5, pp. 691–723, 2011, doi: 10.1080/13647830.2011.558594.

[4] J. E. Shepherd, I. O. Moen, S. B. Murray, and P. A. Thibault, "Analyses of the cellular structure of detonations," *Symposium (International) on Combustion*, vol. 21, no. 1, pp. 1649–1658, 1988, doi: 10.1016/S0082-0784(88)80398-9.

[5] J. J. Lee, D. Garinis, D. L. Frost, J. H. S. Lee, and R. Knystautas, "Two-dimensional autocorrelation function analysis of smoked foil patterns," *Shock Waves*, vol. 5, no. 3, pp. 169–174, 1995, doi: 10.1007/BF01435524.

Water sheet breakup dynamics and spray characterization for confined leading-edge cooling

The advancement of hypersonic aviation is critically associated with refined internal cooling of aerodynamically heated, sharp airfoil leading edges. Liquid sheet generation leads to the formation of stable planar sprays which may optimally address this demand. Liquid sheet jet atomization has been an integral research subject in accelerating controlled spray generation. Herein, anodically bonded silicon-glass wafers with etched convergent micro-nozzles are employed to generate liquid sheets and experimentally investigate their breakup and atomization. High-speed shadowgraph imaging is utilized to visualize the sheet breakup and image processing algorithms to quantify the spray characteristics. The variation of the nozzle thickness, outlet width, and converging angle compose a test matrix of 50 geometries. The investigated geometric-flow ranges correspond to Weber numbers ranging between 10 and 10000.

Qualitative and quantitative sheet and spray characteristics are evaluated. A regime classification distinguishing the generated sheets' morphology differentiates the working conditions for jet impingement, uniform, and multiscale spray cooling. A sub-class leading to a narrow out-of-plane spray is identified. The spray formation -initiated with the liquid sheet formation- was found to be delayed by the increase of the nozzle's outlet and accelerated at larger converging angles. A new correlation for the sheet's breakup length is proposed which principally agrees with analytically and semi-empirically derived formulas in the literature, as the breakup length is a quadratic function of the flow rate. However, the acquired results demonstrate a weak dependence between the breakup length and the nozzle angle. These findings enhance the understanding of water sheet jet disintegration and offer a solid foundation for designing microfluidic systems tailored for precise and planar spray formation.

Manuscript 88 submitted to 64th Israel Annual Conference on Aerospace Sciences (IACAS). Received January 15, 2025.

Characterizing the aerodynamic degradation of a damaged UAV pusher propeller

Elad Bar-On*, Yosef Pikel †, and Hadar Ben-Gida‡
Israeli Air Force, Tel Aviv 6473428, Israel

This study presents a comprehensive evaluation of the aerodynamic performance of a pusher propeller, focusing on comparing intact and damaged (broken) configurations following a previous flight accident. The two propeller configurations were 3D scanned to acquire a high-resolution model. Challenges associated with surface irregularities and alignment during scanning were addressed through advanced contrast techniques and multiple scanning passes. The resulting point-cloud data was processed into solid models using segmentation, mesh fitting, and iterative refinement, ensuring high quality of the final geometry representation. A low-fidelity aerodynamic analysis based on the blade element momentum theory (BEMT) was employed to evaluate the performance of the propeller geometries. The BEMT code from the getPROP-Matlab software was used, and the sectional aerodynamic database was computed from XFOIL for a range of operating conditions. Experiments conducted exclusively on the intact (full) propeller configuration provided fundamental validation of the computational method, demonstrating good agreement between experimental and numerical results. The results highlight the significant impact of structural integrity on aerodynamic performance. The intact (full) configuration consistently demonstrated superior thrust generation and efficiency compared to the damaged (broken) propeller. Additionally, the analysis revealed that the broken configuration required higher RPM for the same thrust, further emphasizing the performance degradation caused by structural damage. In conclusion, this study demonstrates the effectiveness of integrating advanced 3D scanning, solid modeling, and computational tools for accurately evaluating the aerodynamic performance of propeller designs.

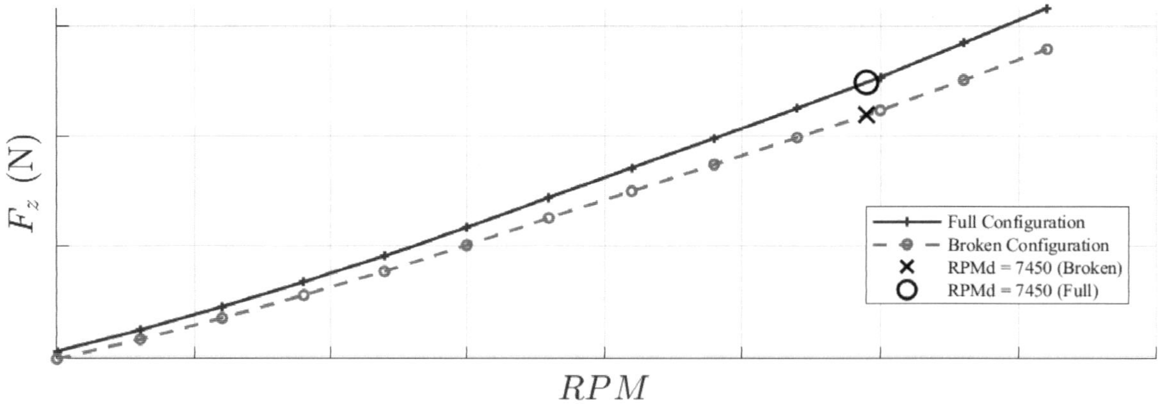

Fig. 1 Comparison of thrust force (F_z) as a function of RPM for intact (full) and damaged (broken) propeller configurations.

*Aerospace Engineer, UAV Propulsion Officer, Branch UAV A, eladb143@gmail.com.
†Head of UAV Propulsion, Branch UAV A.
‡Aerodynamics Subject Matter Expert, Aeronautical Engineering Branch, bengida@idf.il.

Manuscript 68 submitted to 64th Israel Annual Conference on Aerospace Sciences (IACAS). Received January 10, 2025.

Swept and Tapered Wings in Dynamic Motion

Sarah M. Sullivan, Evan Mahns, and Michael Amitay
Rensselaer Polytechnic Institute, Troy, NY, 12180

The objective of the present work is to further investigate experimentally the complex flow separation over swept and tapered finite wings by analyzing the effects of static and dynamic pitch, roll and yaw with varying pitch, roll, and yaw angles. Two types of dynamic motions were explored, sinusoidal and ramp-up/ramp-down motions. The effects of pitch amplitude, rate, and acceleration on the aerodynamic loads were explored. Aerodynamic load measurements were collected to analyze these effects. 3-D flow fields, during dynamic motion, were also obtained using Time Resolved Stereo particle image velocimetry (TR-SPIV). Most of the research was conducted on a finite wing with aspect ratio, $AR = 4$, an airfoil geometry of NACA 0015 and a taper ratio of 0.27. The leading edge of the wing had no sweep and the trailing edge was swept forward 30°. The model was connected to the Motion Position System (MPS) via a sting balance. The MPS allows for motion in pitch, roll and yaw. It was used for static and dynamic load measurements at rotational speeds up to 5 degrees per second in pitch and yaw and 10 degrees per second in roll. For the static experiments, yaw had a larger effect on the aerodynamic loads compared to roll. During dynamic motion, hysteresis was observed during sinusoidal pitch, whereas high frequency transitory behavior was observed during dynamic pitch-up/pitch-down pitch. In the presentation, the effect of pitch, yaw and roll on the aerodynamic loads and the flowfield, in static and dynamic conditions, will be discussed.

Manuscript 25 submitted to 64th Israel Annual Conference on Aerospace Sciences (IACAS). Received December 23, 2024.

Optimizing drones delivery path for low-noise flight

Barak Deutscher*
Technion Israel Institute of Technology, Haifa 3200003, Israel

Oksana Stalnov†
Elbit Systems Ltd., Haifa 3100401, Israel

Hadar Ben-Gida‡
Technion Israel Institute of Technology, Haifa 3200003, Israel

For the past decade, airborne package deliveries in urban and rural environments via drone platforms such as multi-rotors have become a fast-evolving market in general aviation. Despite numerous advantages (e.g., maneuverability, low cost, etc.), multi-rotor platforms result in a significant acoustic pollution penalty produced by the harsh and distinguished sound emitted by the rotors, which is already adversely affecting the health of numerous residential communities. Since drone noise regulations are frequently discussed and updated, a growing need emerges for advanced multi-rotor mission planning tools incorporating acoustic detection constraints. The current study presents a novel mission planning framework to solve the multi-rotor delivery service problem. It combines the multi-rotor performance and acoustics with the navigation algorithm, given acoustic constraints for the surrounding communities, to yield a low-noise signature and optimal delivery route. The mission-planning algorithm employs a global optimization approach, performing an optimal graph search over key points from delivery toward destination, obtained by analyzing the Digital Terrain Map (DTM). The acoustic constraint is defined using the loudness psychoacoustic metric, which allows the implementation of the constraint as a minimal distance between the multi-rotor and the community. By incorporating the acoustic constraint as a geometric constraint on a three-dimensional DTM and using the A-star navigation algorithm, a complete framework for acoustically optimizing the drone flight path is established. The mission planning framework presented herein demands low computational resources and thus has the potential to be integrated into a drone's flight computer for real-time mission planning.

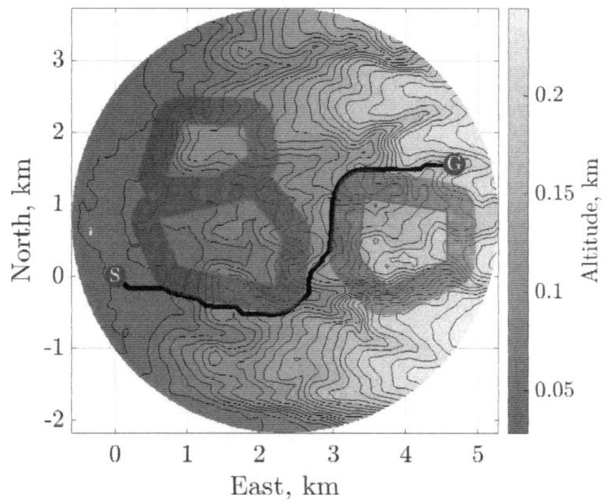

Fig. 1 Optimal drone's flight path given an acoustic detection constraint of 400 m. The 'no-fly zone' under the acoustic constraint is the orange area around the community.

*M.Sc. student, Faculty of Aerospace Engineering, barakde@campus.technion.ac.il.
†Aerodynamic Engineer, Elbit Systems Ltd., miss.stalnov@gmail.com.
‡Research Fellow, Faculty of Aerospace Engineering, bengida@technion.ac.il.

Manuscript 37 submitted to 64th Israel Annual Conference on Aerospace Sciences (IACAS). Received January 3, 2025.

Experimental analysis of a vibro-acoustic response of an airborne structure to boundary layer excitation

Maayan Naschletashvili*
Technion Israel Institute of Technology, Haifa 3200003, Israel

Oksana Stalnov†
Elbit Systems Ltd., Haifa 3100401, Israel

Hadar Ben-Gida‡
Technion Israel Institute of Technology, Haifa 3200003, Israel

Airborne system experiences a range of environmental stresses during its life cycle. Generally, it is essential to ensure those stresses do not affect the system in terms of performance and structural integrity. Vibro-acoustic loads are known as one of the significant stresses that act on an airborne system. Such loads present a major engineering challenge involving a complex interaction between two physical phenomena. The first is the aerodynamic excitation, which includes the interactions between the fluid and the structure due to a turbulent boundary layer (TBL) and corresponding pressure fluctuations. The second phenomenon is structural vibrations, which involve the structure responses and the resulting pressure fluctuations around it. Such fluctuations on the structure's outer surface influence the fluid and change the TBL characteristics. At the same time, the fluctuations are transmitted through the structure and induce a pressure field in the inner cavity of the structure, which causes both noise and vibrations to the inner assemblies. The problem is intrinsically multidisciplinary, encompassing mechanical vibrations, aerodynamics, and both external and internal acoustics. This study aims to conduct interdisciplinary research into the vibro-acoustic response of a flat plate through wind tunnel experiments at the anechoic wind tunnel at the Faculty of Aerospace Engineering at the Technion. The experimental setup included two plates. A flexible thin plate was chosen to measure and characterize the vibro-acoustic response of a structure. On the side of the plate that was not exposed to the flow, an anechoic wooden box was fastened so that no outer noise could penetrate the box beside the plate. A thicker plate equipped with remote microphone probes was used to characterize the pressure fluctuations in space and time. The two plates were positioned as an extension to the wind tunnel nozzle, such that they were flush mounted with the freestream flow. The pressure fluctuations on the thick plate were followed by the structural response of the thin plate, which generated noise inside the wooden box. This study measured and analyzed all those phenomena to enlighten the understanding of the vibro-acoustic problem.

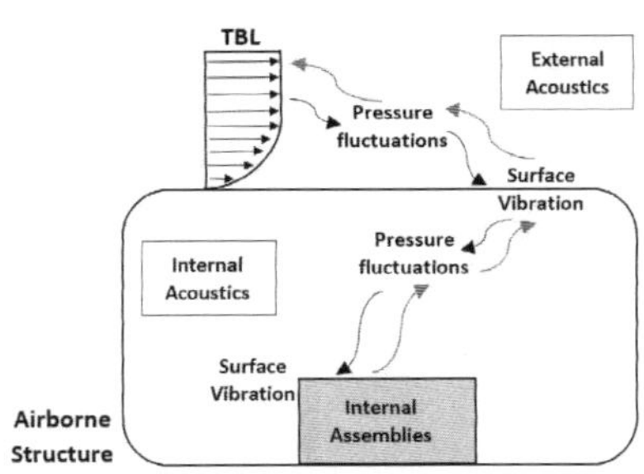

Fig. 1 Illustration of the vibro-acoustic problem

*M.Sc. student, Faculty of Aerospace Engineering, maayanshvili@campus.technion.ac.il.
†Aerodynamic Engineer, Elbit Systems Ltd., miss.stalnov@gmail.com.
‡Research Fellow, Faculty of Aerospace Engineering, bengida@technion.ac.il.

Manuscript 81 submitted to 64th Israel Annual Conference on Aerospace Sciences (IACAS). Received January 13, 2025.

Sustainable Flight and Maneuverability of Guided Projectiles at Transonic Speeds via Circulation Control

Polonsky D., Alon Tzezana G., Dahan A.

IN the modern era of aviation, the design methodology has been greatly influenced by the race to high-speed flight and by the need for reductions in fuel consumption [1]. As a consequence of both the commercial and private sector's demands, solutions for aerial platforms that both facilitate a reduction in power consumption, as well as allow for high maneuverability are on the rise.

Active flow control (AFC), has shown great promise since its early research stages in the the mid 1900's, and currently, the Control of Revolutionary Aircraft with Novel Effectors (CRANE) Project [2] led by DARPA and developed by Lockheed Martin would be the first large demonstrator to maneuver using only fluidics effectors at commercial altitudes and speeds.

Of the different AFC methods, Circulation Control (CC) is a promising candidate. By injecting a high-speed jet around a blunt trailing-edge, centrifugal pressure attributed to the Coandă effect [3], produces an increase in circulation and thus an increase in lift [4], with early research endeavors showing that a substantial lift increase can be achieved [5].

Although originally designed for Short Takeoff and Landing (STOL), later applications started promoting a CC design aimed for maneuvering to either complement or replace standard lifting surfaces, with demonstrators such the BEA Systems MAGMA demonstrator [6] applying the methodology. Utilizing CC for maneuvering applications is preferable, since not only in replacing the lifting surfaces the drag of the system can be dramatically reduced, less jet power is required for the prescribed lift.

Nevertheless, as the Mach number increases, specifically to transonic speeds, the efficacy of CC decreases significantly [7], requiring specialized design to handle jet detachment. Simultaneously, the literature on the subject is lacking with only a few published new wing experiments [8], and more recently in the 2010's, the Fundamental Aerodynamic Subsonic Transonic Modular Active Flow (FAST-MAC) experiments [9]. Difficulties also arise in terms of mechanical complexity. Circulation Control, which requires an air injection system was originally designed for airplanes and helicopters, while fundamentally, projectiles are more compact and thus undoubtedly suffer from implementation challenges.

This study, is then aimed at extending the physical knowledge of using circulation-control at the transonic regime via numerical simulations. Currently, 2D simulations were conducted within the RANS framework, to understand the proper modeling of the boundary conditions, the effects of geometric design and to establish a margin for which CC is optimal in both cruise and maneuvering (with ongoing 3D simulations using both RANS and wall modeled LES), as well as tackling the engineering challenge of design, implementation and testing of a small scale injection system.

Manuscript 54 submitted to 64th Israel Annual Conference on Aerospace Sciences (IACAS). Received January 8, 2025.

References

[1] Ashok, A., "World Economic Forum," , 2023. URL https://www.weforum.org/stories/2023/10/speed-and-sustainable-aviation/.

[2] Montoro, J., Carbaugh, W. L., and Miller, D. N., "DARPA CRANE Program Overview by Lockheed Martin," *AIAA SCITECH 2023 Forum*, American Institute of Aeronautics and Astronautics, 2023. https://doi.org/10.2514/6.2023-1808.

[3] Coanda, H., "Device for Deflecting a Stream of Elastic Fluid Projected into an Elastic Fluid," , 1936. URL https://patents.google.com/patent/US2052869A/en.

[4] Englar, R., "Circulation control pneumatic aerodynamics: blown force and moment augmentation and modification-Past, present and future," *Fluids 2000 Conference and Exhibit*, 2000, p. 2541. https://doi.org/10.2514/6.2000-2541.

[5] Jones, G. S., and Joslin, R. D., *Applications of circulation control technology*, American Institute of Aeronautics and Astronautics, 2006. https://doi.org/10.2514/4.866838.

[6] Warsop, C., and Crowther, W. J., "Fluidic Flow Control Effectors for Flight Control," *AIAA Journal*, Vol. 56, No. 10, 2018, pp. 3808–3824. https://doi.org/10.2514/1.J056787.

[7] Englar, R. J., "Two-Dimensional Transonic Wind Tunnel Tests of Three 15-Percent Thick Circulation Control Airfoils," Tech. rep., Defense Technical Information Center, 1970. https://doi.org/10.21236/AD0882075.

[8] Alexander, M. G., Anders, S. G., Stuart, K., Florance, J. P., and Keller, D. F., "Trailing Edge Blowing on a Two Dimensional Six Percent Thick Elliptical Circulation Control Airfoi up to Transonic Conditions," Tech. rep., NASA Langley Research Center, 2005. URL https://ntrs.nasa.gov/citations/20050165090.

[9] Jones, G. S., Milholen, W. E., Chan, D. T., Allan, B. G., Goodliff, S., Pack Melton, L. G., Anders, S. G., Carter, M. B., and Capone, F., "Development of the circulation control flow scheme used in the NTF semi-span FAST-MAC model," *31st AIAA Applied Aerodynamics Conference*, 2013, p. 3048. https://doi.org/10.2514/6.2013-3048.

Static Aeroelasticity and Stability of Very Flexible Swept Wings

Bar Revivo * and Daniella E. Raveh†
Technion - Israel Institute of Technology, Haifa, Israel

The study investigates the aeroelastic behavior of very flexible, swept-back wings through computational analyses and comparisons with straight-wing models. The study evaluates how large deformations and varying sweep angles affect the static aeroelastic response, natural frequencies, mode shapes, and flutter characteristics, using swept wing models based on the Pazy wing as a benchmark. Structural and aeroelastic analyses were performed with Nastran finite-element software for the linear, small deformations domain. Nonlinear aeroelastic simulations were conducted using the Modal Rotation Method, a modal-based nonlinear analysis framework suitable for beam-like wing geometries that undergo large deformations. Aeroelastic simulations show how bending-torsion coupling in swept wings, which exists even in their undeformed states, affects the stability and static behavior of swept wings compared to straight wings.

Flutter analysis was performed using a modified p-k method, in which the generalized mass and unsteady aerodynamic matrices were reevaluated about the statically deformed shape corresponding to each angle of attack and airspeed combination. These analyses yield the correct trends for frequencies and damping values with varying airspeed for straight and swept wings that undergo large deformations. The hump flutter mode, a characteristic of the straight Pazy wing, was found to be stable in the swept wings, across all studied configurations. The only flutter mechanism identified in the swept wings was a higher-speed mechanism, combining first and second out-of-plane bending and first torsion. Normalizing the flutter characteristics revealed that the flutter frequency, velocity, and flutter mechanism follow a similar trend regardless of sweep angle and that, therefore, flutter prediction based on tip deflection might be possible.

Static aeroelastic analyses indicated that while swept-back wings tend to twist nose-down and limit wing deformations, significant deformations can still occur even at high sweep angles. Additionally, the results suggest that a strip-theory aerodynamic model, which assumes undeformed lift-line slopes, may not be adequate to predict flexible swept wing deformations, even at small deformations. The study highlights the need for nonlinear structural modeling and high-fidelity aerodynamic modeling in flexible swept wings and suggests further experimental validation for improved aeroelastic performance predictions.

*Graduate Student, Faculty of Mechanical Engineering, Technion, bar.revivo@campus.technion.ac.il
†Professor, Faculty of Aerospace Engineering, Technion

Manuscript 2 submitted to 64th Israel Annual Conference on Aerospace Sciences (IACAS). Received December 10, 2024.

Aeroelasticity of Plates in Supersonic Channel Flow

Maxim Freydin *
Technion – IIT, Haifa, Israel, 3200003

Abstract

Unsteady aerodynamic models used to investigate the aeroelasticity of plates can be classified into three levels of complexity which affects accuracy and computational burden. The first level is Piston Theory aerodynamics which is linear and local in space but simple mathematically. The second level is the linear Potential Flow (PF) aerodynamics which accounts for non-local effects through convolution terms [1]. The third level involves computational fluid dynamics (in its various forms) the use of which is not practical when the parameter space is very large. The present work deals with elastic plates in supersonic channel flow where reflections from bounding walls might affect the aeroelastic response. This configuration is of interest for the wide range of wind tunnel experiments conducted in the context of aero-thermal-elasticity of plates [2, 3]. The non-local effects introduced by bounding walls make linear PF aerodynamics the simplest computationally feasible model to analyze the fluid-structure coupling over a broad parameter range. However existing solutions assume unbounded freeflow and use integral transforms to obtain the solution in modal coordinates [1]. In this work a PF solution is obtained for the wall-bounded case (channel flow) in both physical and modal coordinates and in frequency and time domains. The solution is obtained by modifying the spanwise integral transform to prescribe impermeable boundary conditions at the bounding walls. Figure 1 illustrates the pressure perturbation distribution on the plate surface at multiple time steps following an impulse displacement in the first structural mode. In this example a bounding wall is placed on one side of the plate and reflections are observed. The reflections break the span-wise symmetry of the unbounded case and might lead to aeroelastic coupling between structural modes that are nominally not coupled. We investigate how the aeroelastic system is influenced by key parameters, including the plate's proximity to the walls, aspect ratio, and Mach number.

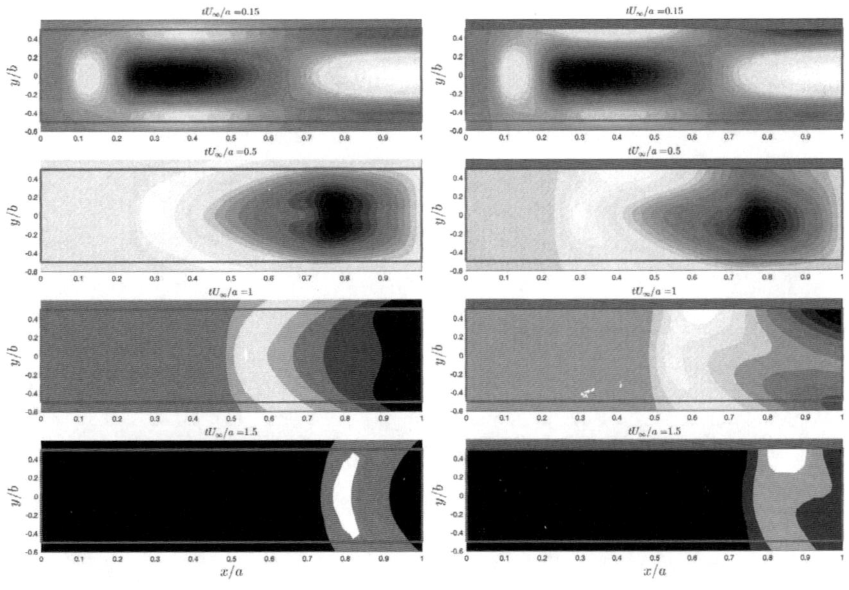

Fig. 1 Unsteady pressure impulse response to displacement in the first structural mode at Mach 2. The left column shows the case of freestream flow (no bounding walls). The right column shows the case with a bounding wall (solid blue) on one side. The rows show the response evolution in time.

*Lecturer, Aerospace Engineering, freydinmaxim@technion.ac.il

Manuscript 13 submitted to 64th Israel Annual Conference on
Aerospace Sciences (IACAS). Received December 17, 2024.

References

[1] Dowell, E. H., *Aeroelasticity of plates and shells*, Noordhoff International Publishing (now Springer), 1974.

[2] Spottswood, S. M., Beberniss, T. J., Eason, T. G., Perez, R. A., Donbar, J. M., Ehrhardt, D. A., and Riley, Z. B., "Exploring the response of a thin, flexible panel to shock-turbulent boundary-layer interactions," *Journal of Sound and Vibration*, Vol. 443, 2019, pp. 74 – 89. https://doi.org/https://doi.org/10.1016/j.jsv.2018.11.035.

[3] Brouwer, K. R., Perez, R. A., Beberniss, T. J., Spottswood, S. M., and Ehrhardt, D. A., "Experiments on a Thin Panel Excited by Turbulent Flow and Shock/Boundary-Layer Interactions," *AIAA Journal*, Vol. 59, No. 7, 2021, pp. 2737–2752. https://doi.org/10.2514/1.J060114.

Manuscript 13 submitted to 64th Israel Annual Conference on Aerospace Sciences (IACAS). Received December 17, 2024.

Experimental Investigation of the F-16 Shock Buffet with Comparison to Flight Tests

Maj. Tzlil Nahom Jidovetski* and Lt. Col. Michael Iovnovich† and Daniella E. Raveh‡

I Introduction

Buffet is an unsteady aeromechanical phenomenon where flow instabilities around an aircraft's lifting surfaces cause oscillatory forces, which may degrade the platform handling qualities and structural integrity characteristics. While buffet originates from a flow phenomenon, its characteristic frequencies often fall within the range of an aircraft's structural dynamics natural frequencies, which may lead to dynamic loads magnification by means of aeroelastic resonance interaction. Consequently, buffet is a significant design factor and certification challenge in flight envelope expansion, structure strength substantiation, and fatigue life assessment processes for both commercial and fighter platforms.

This study focuses on transonic buffet (shock-buffet), characterized by self-sustained shock-wave oscillations caused by shock-boundary layer interactions at certain Mach numbers and angles of attack. This phenomenon has been extensively investigated in the literature, both numerically [1–11] and in wind tunnel experiments [12–22] since the 1980s. Previous shock-buffet studies on fighter jets such as the F-104 [23] and F-111 [24] are very limited and mainly based on instrumented flight test measurements. The F-35 exhibits multiple buffet phenomena, including transonic wing buffet, which significantly impacted its structural design and certification processes [25]. Predicting and quantifying buffet loads are essential for the structural design purposes of such configurations.

Reliable buffet load models are often available only late in the design process, based on costly flight tests. Therefore, there is a strong motivation to better understand transonic buffet on fighter aircraft and develop efficient assessment strategies using CFD and wind tunnel tests.

II Work Description

This study is conducted as part of a multi-year effort by the IAF to develop and validate a buffet-load prediction capability primarily for the purpose of enhancing fighter aircraft external store certification processes. The current study is focused on predicting F-16 transonic buffet loads based on wind-tunnel test data. The test was conducted on a rigid F-16 model (1:15) instrumented with unsteady pressure transducers at transonic flow conditions. The first goal of this test was to identify and map the F-16 buffet envelope in terms of Mach number and angles of attack. Once buffet conditions were identified, unsteady pressure data was collected to characterize the phenomenon and compare its physical mechanisms and characteristics with those previously known from transport

*Head of Flight Sciences Group, tzlilnahom@gmail.com
†Lead Aeroelasticity Engineer, Flight Sciences Group, michaeliovnovich@idf.il
‡Professor, Faculty of Aerospace Engineering, daniella@technion.ac.il

Manuscript 51 submitted to 64th Israel Annual Conference on Aerospace Sciences (IACAS). Received January 7, 2025.

aircraft buffet studies. A comparative study was then conducted for various external store loadings to assess the effects of external stores at different store stations on wing buffet characteristics. The study concludes with a comparison of wind-tunnel results against available flight test data. The findings are expected to suggest prediction strategies for assessing the effects of external stores, as well as to enhance the understanding and prediction of buffet loads based on wind tunnel test data

III Preliminary Results

Preliminary results are presented for both clean store stations and loaded store stations (with missiles/stores) configurations. Figure 1 compares the buffet intensity envelopes for both configurations. The buffet intensity was calculated by integrating the RMS levels of the pressure coefficient fluctuation on the upper surface of the left wing under varied flow conditions, including Mach number and angle of attack. The buffet intensity levels are normalized by the maximum intensity from both configurations for comparison. Both configurations exhibit distinct buffet intensity peaks centered around Mach 0.83. The fully loaded wing configuration shows an additional high-intensity region at AOA of 12° - 14° across all Mach numbers. Additionally, the buffet envelope is significantly wider for the loaded wing configuration than for the clean wing configuration.

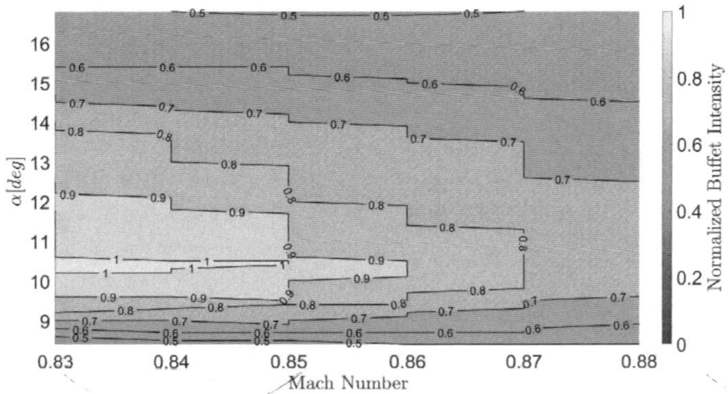

(a) Unloaded wing configuration

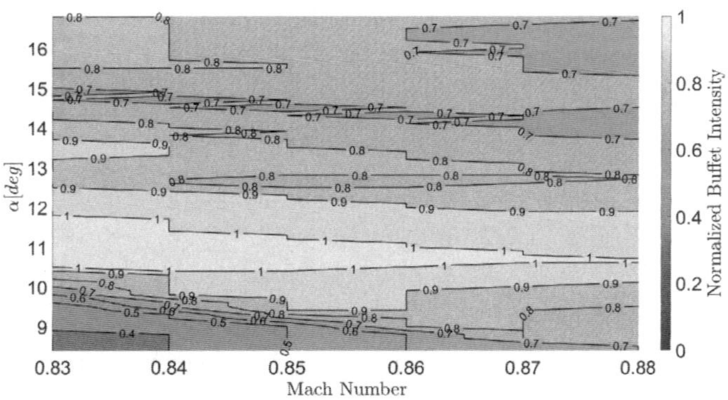

(b) Loaded wing configuration

Figure 1. Buffet Intensity envelope variation with Mach number and angle of attack

Manuscript 51 submitted to 64th Israel Annual Conference on Aerospace Sciences (IACAS). Received January 7, 2025.

References

[1] Iovnovich, M. and Raveh, D. E., "Numerical Study of Shock Buffet on Three-Dimensional Wings," *AIAA Journal*, Vol. 53, No. 2, 2015, pp. 449–463.

[2] Iovnovich, M. and Raveh, D. E., "Reynolds-Averaged Navier-Stokes Study of the Shock-Buffet Instability Mechanism," *AIAA journal*, Vol. 50, No. 4, 2012, pp. 880–890.

[3] Lee, B., "Self-Sustained Shock Oscillations on Airfoils at Transonic Speeds," *Progress in Aerospace Sciences*, Vol. 37, No. 2, 2001, pp. 147–196.

[4] Poplingher, L. and Raveh, D. E., "Comparative Modal Study of the Two-Dimensional and Three-Dimensional Transonic Shock Buffet," *AIAA Journal*, Vol. 61, No. 1, 2023, pp. 125–144.

[5] Raghunathan, S., Gillan, M., Cooper, R., Mitchell, R., and Cole, J., "Shock Oscillations on Biconvex Aerofoils," *Aerospace Science and Technology*, Vol. 3, No. 1, 1999, pp. 1–9.

[6] Deck, S., "Numerical Simulation of Transonic Buffet Over a Supercritical Airfoil," *AIAA journal*, Vol. 43, No. 7, 2005, pp. 1556–1566.

[7] Thiery, M. and Coustols, E., "URANS Computations of Shock-Induced Oscillations Over 2D Rigid Airfoils: Influence of Test Section Geometry," *Flow, Turbulence and Combustion*, Vol. 74, No. 4, 2005, pp. 331–354.

[8] Goncalves, E. and Houdeville, R., "Turbulence Model and Numerical Scheme Assessment for Buffet Computations," *International Journal for Numerical Methods in Fluids*, Vol. 46, No. 11, 2004, pp. 1127–1152.

[9] Sartor, F. and Timme, S., "Reynolds-averaged Navier-Stokes simulations of shock buffet on half wing-body configuration," in "53rd AIAA Aerospace Sciences Meeting," , 2015, p. 1939.

[10] Plante, F., Dandois, J., and Laurendeau, É., "ZDES and URANS Simulations of 3D Transonic Buffet Over Infinite Swept Wings," in "Progress in Hybrid RANS-LES Modelling," Springer, pp. 259–269, 2020.

[11] Timme, S., "Global shock buffet instability on NASA common research model," in "AIAA Scitech 2019 Forum," , 2019, p. 0037.

[12] McDevitt, J. B., Levy Jr, L. L., and Deiwert, G. S., "Transonic Flow About a Thick Circular-Arc Airfoil," *AIAA Journal*, Vol. 14, No. 5, 1976, pp. 606–613.

[13] McDevitt, J. B. and Okuno, A. F., "Static and Dynamic Pressure Measurements on a NACA 0012 Airfoil in the Ames High Reynolds Number Facility," Tech. rep., NASA TR-TP-2485, 1985.

[14] Jacquin, L., Molton, P., Deck, S., Maury, B., and Soulevant, D., "Experimental Study of Shock Oscillation Over a Transonic Supercritical Profile," *AIAA journal*, Vol. 47, No. 9, 2009, pp. 1985–1994.

[15] Dandois, J., "Experimental Study of Transonic Buffet Phenomenon on a 3D Swept Wing," *Physics of Fluids*, Vol. 28, No. 1, 2016, p. 016101.

[16] Benoit, B. and Legrain, I., "Buffeting Prediction for Transport Aircraft Applications Based on Unsteady Pressure Measurements," in "5th Applied Aerodynamics Conference," , 1987, p. 2356.

[17] Koike, S., Ueno, M., Nakakita, K., and Hashimoto, A., "Unsteady pressure measurement of transonic buffet on NASA common research model," in "34th AIAA Applied Aerodynamics Conference," , 2016, p. 4044.

[18] Lee, B., Ellis, F., and Bureau, J., "Investigation of the Buffet Characteristics of Two Supercritical Airfoils," *Journal of aircraft*, Vol. 26, No. 8, 1989, pp. 731–736.

[19] Paladini, E., Dandois, J., Sipp, D., and Robinet, J.-C., "Analysis and comparison of transonic buffet phenomenon over several three-dimensional wings," *AIAA Journal*, Vol. 57, No. 1, 2019, pp. 379–396.

[20] Molton, P., Dandois, J., Lepage, A., Brunet, V., and Bur, R., "Control of Buffet Phenomenon on a Transonic Swept Wing," *AIAA Journal*, Vol. 51, No. 4, 2013, pp. 761–772, doi:10.2514/1.J051000.

[21] Masini, L., Timme, S., and Peace, A. J., "Analysis of a civil aircraft wing transonic shock buffet experiment," *Journal of Fluid Mechanics*, Vol. 884, 2020, p. A1, doi: 10.1017/jfm.2019.906.

[22] Sugioka, Y., Koike, S., Nakakita, K., Numata, D., Nonomura, T., and Asai, K., "Experimental analysis of transonic buffet on a 3D swept wing using fast-response pressure-sensitive paint," *Experiments in Fluids*, Vol. 59, 2018, pp. 1–20.

[23] Friend, E. L. and Sefic, W. J., "Flight Measurements of Buffet Characteristics of the F-104 Airplane for Selected Wing-Flap Deflections," Tech. rep., NASA TN D-6943, 1972.

[24] Monaghan, R. C., "Flight-Measured Buffet Characteristics of a Supercritical Wing and a Conventional Wing on a Variable-Sweep Airplane," Tech. rep., NASA-TP-1244, 1978.

[25] Ellis, R. M., Gross, P., Yates, J. B., Casement, J. R., Chichester, R. H., and Nesmith, K., "F-35 Structural Design, Development, and Verification," in "2018 Aviation Technology, Integration, and Operations Conference," , 2018, p. 3515.

Manuscript 51 submitted to 64th Israel Annual Conference on Aerospace Sciences (IACAS). Received January 7, 2025.

Aeroelastic Coupling of Hard Maneuvering Aircraft

Dor Naftaly* and Daniella E. Raveh†
Technion - IIT, Haifa 3200000, Israel

Aircraft structures are inherently elastic, resulting in a coupling between flight dynamics, structural dynamics, and aerodynamics. Yet, the common approach is to solve the flight dynamics and aeroelastic problems independently, only accounting for the mutual aerodynamic influences. The current study aims to bridge this gap by deriving the coupled flight-mechanics-aeroelastic equations of motion (EOM) and using them to examine the inertial coupling between flight dynamics and aeroelasticity during high-load maneuvers, with a particular focus on evaluating the impact of maneuver load factors on flutter onset.

The main rigid-body elastic couplings are the aerodynamic coupling and the inertial coupling, and their significance depends on the vehicle's flexibility (i.e., typical linear flexibility or highly flexible configuration) and the maneuver's intensity (i.e., light or hard maneuver). For small deformations and light maneuvering cases, the inertial coupling is negligible compared to the aerodynamic coupling, as demonstrated, for example, by Francesco *et al.* [1]. However, for very flexible configurations or during hard maneuvers, the inertial coupling is non-negligible, calling for a coupled flight-mechanics aeroelastic solution. During hard maneuvering, the rigid and elastic EOM are coupled through the (non-negligible) angular rates, even in mildly flexible configurations where the elastic deformations are small.

Inertial aeroelastic coupling is typically not considered in the certification flutter analyses, although it might impact the stability envelope of hard-maneuvering aircraft such as fighter jets, that can maintain steady, level turns with high load factors for extended periods. The current study includes the derivation of the fully coupled flight dynamics and aeroelastic EOM, and stability analysis accounting for the coupling terms. The test case presents the impact of these couplings on the flutter envelope of an F-16 model through a survey of 2016 different store configurations. Several store configurations characterized by notable flutter mechanisms, such as hard flutter and hump mode instabilities, for which the significance of the inertial coupling effect is expected to vary, are investigated.

Of all inertial coupling terms, the most significant one was found to be the term that couples rigid rates in the nominal state to the modal rigid and elastic motion. In some cases, introducing this term resulted in up to a 9% increase or a 2% decrease in flutter velocity. From the full survey, three specific cases were selected for further investigation to study the dependency of flutter onset speed on the load factor, sensitivity to numerical parameters, and the importance of accurately solving the trimmed state. As an example, Figure. 1 shows the variation of the aeroelastic frequencies and damping values versus normalized airspeed for one of the studied cases, computed at a straight and level flight and about a symmetric coordinated turn of load-factor five. For this case, the flutter speed in the turn increases by 9% compared to straight and level flight.

The results demonstrate that the load factor has an approximately linear effect on flutter onset speed. Future work will further explore the impact of rigid-elastic coupling terms in the linearized coupled EOM in modal coordinates on flutter characteristics and dynamic response, focusing on the significance of the different coupling terms in different applications.

*Ph.D. Student, Faculty of Aerospace Engineering, dor.naf01@gmail.com
†Professor, Faculty of Aerospace Engineering

Manuscript 52 submitted to 64th Israel Annual Conference on Aerospace Sciences (IACAS). Received January 7, 2025.

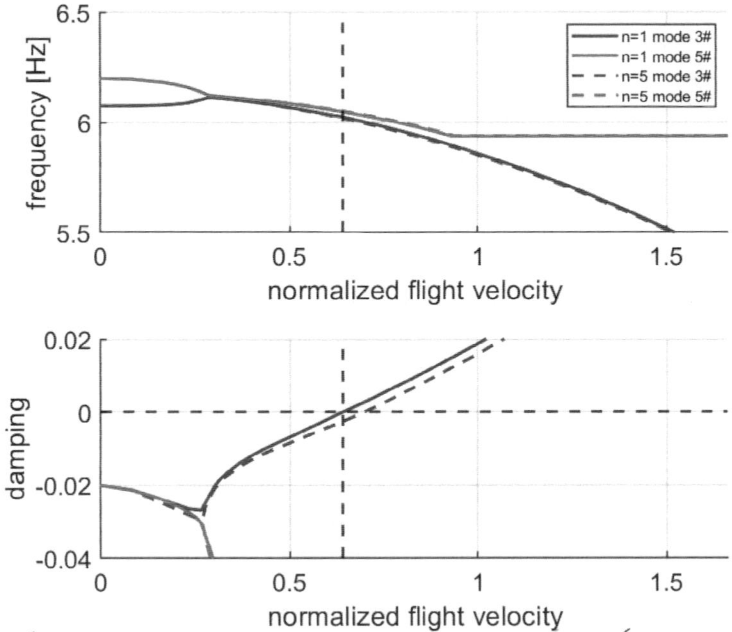

Fig. 1 $\omega V g$ **plot of case 754**

Bibliography

[1] Saltari, F., Riso, C., Matteis, G. D., and Mastroddi, F., "Finite-element-based modeling for flight dynamics and aeroelasticity of flexible aircraft," *Journal of Aircraft*, Vol. 54, No. 6, 2017, pp. 2350–2366.

Numerical Simulation of Flutter Flight Tests Using the Parametric Flutter Margin Method

Guy Gordon-Biberstein [1] and Moti Karpel [2]

[1,2] *Technion – Israel Institute of Technology, 32000 Haifa, Israel*

Abstract

Following previous flutter flight-test conducted at Ohio State University using the Parametric Flutter Margin (PFM) method, numerical simulations were performed for exploring the behavior of the Unattended Aereal Vehicle (UAV) and the data analysis process using the numerical application of PFM by employing a finite-element structural model, verified in ground vibration tests, and common aeroelastic-response methods. Frequency response functions to excitation by added shakers were calculated and used for the identification of the flutter characteristics. The long-term aim is to reduce the uncertainty of flight testing near or at the critical flutter speed, while also improving overall safety and economizing the testing process. The paper concentrates on establishing how effectively the PFM approach, in conjunction with aerodynamic modeling and structural refinement, can capture the actual flutter characteristics. The analysis focuses on identifying the dominant modes that define the flutter onset and stability margins. The flights, performed at elevated speeds with a unique configuration of wing tip mounted shakers with electromagnetically moved masses, provided data that serves as a reference against which numerical models are compared and as a record of the actual aeroelastic behavior. By carefully tuning modal parameters and refining our numerical approach, we aim to enhance the overall accuracy of the predictions. The PFM method, when applied with the static shaker-house masses serving as flutter parameters while the moving masses provide the excitation forces, effectively postpones the flutter onset, allowing safe exprimental access to conditions close to the critical flutter boundary without surpassing it. The simulation results are being used for planning complementary flight-test campaigns with a reduced number of required data points, while mitigating flight risk.

I. Introduction

Flutter is a dangerous aeroelastic instability that arises from the coupling of aerodynamic, structural, and inertial forces. For high-performance or lightweight aircraft, including Unmanned Aerial Vehicles (UAVs), the dynamic interaction of unsteady aerodynamic pressures with flexible airframes can escalate into catastrophic oscillations if not correctly identified and mitigated. Approaching flutter physically in flight is risky, so specialized test methods such as the PFM method [1,2,3] have been proposed and demonstrated to augment stability and reduce real-flight hazards.

The present study builds upon an initial effort started in 2021. Multiple configurations were tested, including different shaker and battery placements in wing-tip pods, seeking to align the predicted flutter speed with the UAV's flight envelope. Early on, the team discovered that certain model assumptions placed the flutter boundary well above the vehicle's maximum airspeed (~35 m/s) in level flight, making it impossible to observe flutter. A second flight campaign introduced extended pods and a re-positioning of mass to bring the boundary nearer 40 m/s, resulting in a dive to reach 42.5 m/s that showed near-flutter phenomena associated with an anti-symmetric torsion-roll mechanism.

The flight data, however, revealed limited correlation with initial Ground Vibration Test (GVT) frequencies. Consequently, an updated finite element (FE) model was needed, factoring in fresh GVT measurements and adopting a g=0.04 structural damping values for the fundamental wing-bending modes and 0.06 for the wing-torsion modes. The collaboration used advanced industrial tools (ZAERO, DynResp) to combine aerodynamic panel solutions with forced-response analyses. The PFM approach then served as a risk-averse scheme to identify the no-shaker flutter boundary while physically never surpassing it. Since certain regulatory frameworks (FAA guidelines) require numerical demonstration of 20 % flutter margin in order to skip test flights, the ability to shift the flutter boundary artificially or prove it is beyond normal flight conditions while keeping the platform safe is immensely beneficial, saving time, cost, and potential damage.

The objectives in this paper are to (a) outline how the final GVT-calibrated UAV model was achieved, (b) describe the PFM approach in detail, including relevant equations, (c) compare ZAERO and DynResp predictions of nominal vs. modified flutter speeds, and (d) overlay key flight test FRFs at 42.5 m/s with numerical forced-response data. This

[1] Graduate student, Faculty of Aerospace Engineering, guygor@campus.technion.ac.il

[2] Professor Emeritus, Faculty of Aerospace Engineering, AIAA Fellow, Karpel@technion.ac.il

work will also show how different window/overlap settings used in the transfer function estimator can obscure the flutter indication. In so doing, we highlight a robust methodology for UAV flutter testing that integrates conceptual design refinements, GVT correlation, advanced simulations, and parametric analysis to ensure safety.

II. Theoretical Framework

A. Fundamentals of the Parametric Flutter Margin (PFM)

The PFM method was developed to enable safer, more precise flutter testing by augmenting the system's stability. Stabilizing elements, such as electromagnetic shakers with known weights can artificially increase the flutter velocity, allowing flight tests to be conducted within a higher stability margin [1,3]. When the "stabilizing parameter" p_f is introduced as a concentrated mass or as damping element, the effective flutter speed rises, reducing the immediate flight risk. In post-processing, one removes the stabilizing effect to infer where the nominal system would actually become unstable. This approach contrasts with traditional direct flutter tests, which often require pushing the flight envelope until large, potentially destructive oscillations occur. To illustrate mathematically, consider the linear homogeneous frequency-domain equation of motion:

$$[A(i\omega)]x(i\omega) = 0, \tag{1}$$

where $\{x(i\omega)\}$ is the state vector combining structural modal displacements, actuator states, and possibly aerodynamic states, while $A(i\omega)$ is the closed-loop system matrix. Conventional flutter solutions (like ZAERO's g-method [6]) look for the velocity–frequency pair (V, ω) that satisfies $\det[A(i\omega)] = 0$. The PFM method adds a known stabilizing parameter p_f to shift the boundary. In the single-input–single-output (SISO) formulation, this modifies Eq. (1) into

$$[A(i\omega) + B_f p_f C_f(i\omega)]\{x(i\omega)\} = \{B_f\} u_f(i\omega), \tag{2}$$

$$y_f = [C_f(iw)]\{x(i\omega)\}$$

where p_f is a single presumably flutter delay parameter, such an added mass or a modal damping term, and B_f, $C_f(i\omega)$ are distribution vectors coupling that parameter to the states. A known sinusoidal or broadband force $u_f(i\omega)$ is applied (e.g. a wing-tip pod shaker), and the resulting response $\{y_f(i\omega)\}$ is measured. By examining the frequency response function (FRF) ratio $\frac{y_f(i\omega)}{u_f(i\omega)}$, one identifies the velocity at which the system would go unstable if p_f were removed:

1. Phase Crossovers: The FRF's phase might pass through 0°, indicating a potential closed-loop instability.
2. Gain Margins: The amplitude of the FRF at the phase-crossover frequency can be viewed as Δp_f, the incremental parameter needed to move the system from stable to unstable.

In many implementations, p_f is a forward mass whose vertical motion influences a critical torsional mode. If removing that mass in analysis yields a 0 dB margin at some velocity V_f, it implies the nominal system would flutter at V_f. The advantage is that the actual flight test can be conducted with the stabilizing mass in place, never physically encountering the true unstable condition. Additional design or test constraints can also be probed by this parametric variation.

B. Linear SISO-PFM Equations

Suppose we define a single output $y_f(i\omega)$ and a single input force $u_f(i\omega)$. If a loop closure is hypothesized:

$$u_f(i\omega) = p_f\, y_f(i\omega), \tag{3}$$

Eq. (2) becomes the homogeneous one in Eq. (1) with $\det[A(i\omega)] = 0$, which implies that the solution $\{x(iw_f)\}$ is the nontrivial flutter mode. For flight data, one does not forcibly close that loop in reality. Instead, a parametric margin analysis in post-processing sees how the measured FRF crosses certain amplitude and phase thresholds. The FRF can be expressed as:

$$G(\omega) = 20\log_{10}|p_f\, y_f(\omega)|, \quad \Phi(\omega) = \angle\big(p_f\, y_f(\omega)\big) \tag{4}$$

A 0 dB margin at $\Phi(\omega_{co}) = 0$ typically designates impending flutter. Alternatively, an incremental flutter parameter Δp_f is often introduced:

$$\Delta p_f = p_f - \frac{u_f(\omega_{co})}{y_f(\omega_{co})}, \tag{5}$$

signifying how much mass or damping "removal" from the stable configuration would drive the system to flutter. The main point is that one can remain safely below the critical speed in flight by retaining p_f. After the flight, the test data is post-processed to see at what speed the system would have gone unstable with p_f factored out.

C. FRF Computation: Flight vs. Numerical Data

The PFM method has been utilized in the DynResp commercial code. Because we want to accurately compare real-world wing-pod accelerometer data recorded during flight with the numerical forced-response solutions obtained from DynResp, we use frequency-response functions as a common framework. The flight data, captured originally in the time domain, are transformed into the frequency domain by applying FFT-based methods (including specific window lengths and overlaps) to the measured input and output (moving-mass and static-mass shaker accelerations). This approach yields a transfer function that we can place on the same Bode plot as the numerically generated FRFs, ensuring direct comparisons of amplitude, phase, and zero-phase crossing for flutter identification. In practice, although two shakers (left and right pods) and multiple sensors are present, we break the problem into repeated SISO loops, with the symmetric measured force or acceleration acting as the "single input" and the anti-symmetric measurement serving as the "single output." Symmetrical versus anti-symmetrical modes are constructed by combining left and right signals, such that

$$y_{\text{sym}}(t) = \frac{y_{\text{right}}(t)+y_{\text{left}}(t)}{2}, \quad y_{\text{anti}}(t) = \frac{y_{\text{right}}(t)-y_{\text{left}}(t)}{2}. \tag{6}$$

Regardless of whether the data comes from flight or numerical simulations, the FRF is defined as

$$\text{FRF}(j\omega) = \frac{\text{output}(j\omega)}{\text{input}(j\omega)}, \tag{7}$$

allowing comparisons on a shared frequency axis. Discrepancies may still arise when flight data are noisy or when window–overlap choices affect phase estimates. On the numerical side, we combine a finite-element structural model (incorporating the "stationary" shaker mass $M_{st} = 0.085$ kg and "moving" mass $M_{mv} = 0.072$ kg) with ZAERO's unsteady aerodynamic matrices, letting DynResp solve forced harmonic responses at each velocity–frequency pair. Because DynResp works natively in the frequency domain, no short-time windowing or overlap is required. The code directly outputs the real and imaginary parts of each sensor's response, $Re[Y]$, $Im[Y]$. From these components, we compute the magnitude and phase via:

$$|Y(v_i, f_k)| = \sqrt{\text{Re}[Y]^2 + \text{Im}[Y]^2}, \quad \phi(v_i, f_k) = arctan\left(\frac{Im[Y]}{Re[Y]}\right). \tag{8}$$

We can also rescale the FRF by including $M_{st}$to examine how adding or removing mass affects the flutter boundary, using:

$$T_{\text{sym}}(j\omega) = \frac{M_{st} \cdot y_{\text{sym}}(j\omega)}{U_{\text{sym}}(j\omega)}, \quad T_{\text{asym}}(j\omega) = \frac{M_{st} \cdot y_{\text{asym}}(j\omega)}{U_{\text{asym}}(j\omega)}. \tag{9}$$

By unifying these data processing steps the flight test accelerometers data turned into FRF estimation via transfer-function-estimator ("tfestimate"), and the numerical FRF based upon DynResp direct force analysis we can overlay both sets of results on the same plots, thereby providing a consistent basis for evaluating aeroelastic stability and detecting flutter.

III. Model Updates and GVT-Calibrated Changes

A. Motivation and Update Procedure

An initial finite-element model had been constructed and partially correlated with GVT data, but flight tests exposed further misalignments, particularly regarding torsional modes. Additionally, the team wanted to lower the flutter speed from an unachievable high 50+ m/s to a safer 35–45 m/s that is in the UAV flight envelope. The second flight campaign managed to reach 42.5 m/s by diving, capturing near-flutter but still leaving some mismatch in normal modes and predicted flutter frequency. To reconcile these, we performed a final GVT calibration step, adopting 4 % structural damping, refined mass distributions (85 g static mass in the pod, plus 72 g for the moving mass of the shaker, 220 g ballast at each rear tip of the pod, etc.), and an extended 6-inch aft pod geometry that further shifts the effective center of gravity behind the elastic axis to lower the flutter velocity. Figure 1 displays the Nastran FEM in Femap.

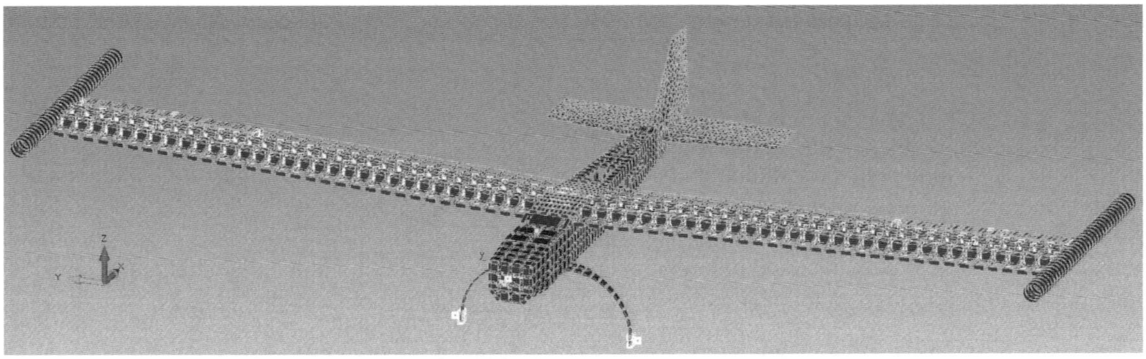

Figure 1. Detailed NASTRAN finite-element model in FEMAP.

Table 1 below summarizes key modifications. The main goal was (1) matching GVT-measured torsion frequencies, (2) achieving accurate weigh-ins for forward CG shift, and (3) ensuring the resulting flutter speed lies near 40–45 m/s. Another factor was clarifying the difference between a "Nominal" model (no shaker static or moving mass) and a "Modified" one (85 g static mass included, without the 72 g of the moving mass of the shaker). By adopting this approach, we differentiate the pure airframe from the actual flight hardware scenario.

Update Aspect	Rationale	Implementation Details
Material G (6061-T6)	Match GVT's higher torsional frequencies.	Shear modulus from 0.31 GPa to 13 GPa
CG X-shift (+10 mm)	Real weighing vs. old estimates	Reassign nodal masses in forward fuselage by 10 mm shift
Pod Extension (6 in.)	Lower torsion/bending frequencies	Extends geometry behind elastic axis in each tip
Li-Po Battery	Accurate real-component weight	Changed 105 g → 108 g (+2.9 %)
Static Mass 220 g/pod	Further reducing flutter speed	Added at the pod's aft tip to shift torsional frequency 5 %
Nylon12 Shear Modulus	GVT indicated underpredicted torsion	Changed 815 MPa → 610 MPa (–25 %)
Total Weight: 13.98 kg	Minor structural re-optimization	Post-integration checks confirm mass distribution & CG location

Table 1. Key Changes in the Updated Model.

The final outcome yields a UAV mass of 13.98 kg, slightly forward CG relative to the older model, and a 6-inch aft extension in each wing-tip pod carrying 220 g near the rear wing tip pod. Together, these changes bring the fundamental torsion mode into better alignment with GVT data (4–10 Hz range) and lower the predicted flutter to around 44 m/s, a speed the UAV can approach in a dive.

B. Updated Normal Modes and Damping

After finalizing the changes in Table 1, we recalculated the first 25 normal modes with NASTRAN. GVT correlation suggests the first torsional mode is within 1–2 % of the measured frequency, while the first bending modes are within 3–4 %. A uniform 4 % structural damping is assumed for the entire UAV. In practice, we generate the .f06 normal modes for subsequent ZAERO and DynResp aeroelastic analyses. We define two principal FE configurations:

- **Nominal Case:** Excluding the shaker's static and moving masses, but including the batteries, wiring and control panel.
- **Modified Case:** Including 85 g static mass in each tip pod, excluding the 72 g moving mass. The "moving mass" parameter is left out of the FE model but will be introduced as p_f in DynResp.

Table 2 shows the different wing-tip pod configurations used in the work. The masses are as follows: (1) 72-gr shaker which is the moving mass (2) the 85-gr shaker house static mass (3) 108 gr batteries (4) the additional Static

Mass 220 g weight. The shaker's static and moving masses are positioned together at the same node. The first 4 elastic natural frequencies and associated mode shapes of the modified configuration are presented in Figure 2, after applying the model updates described earlier.

Nominal	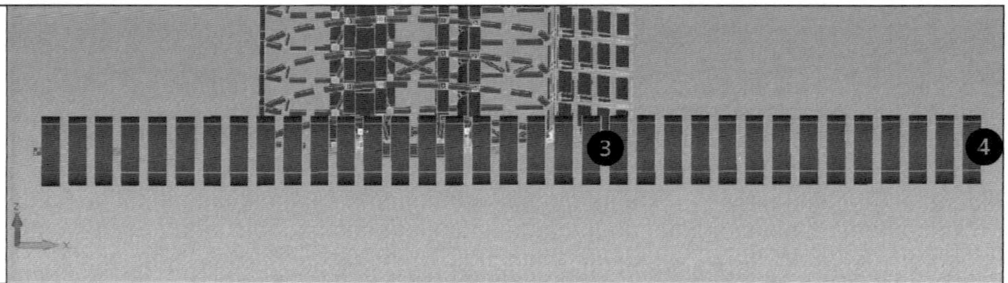
Modified	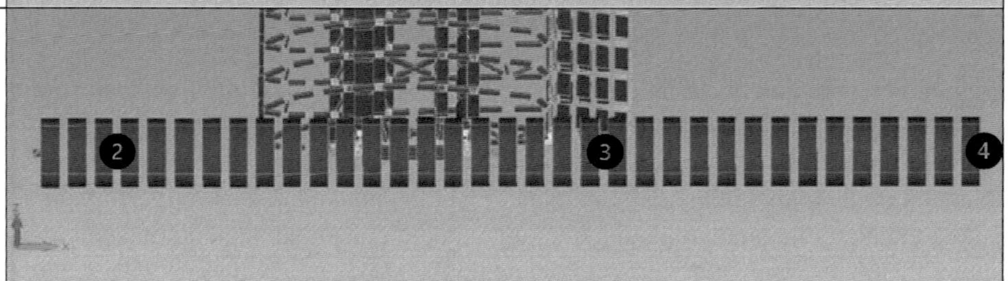
Physical Configuration **(Current Flight Test Payload)**	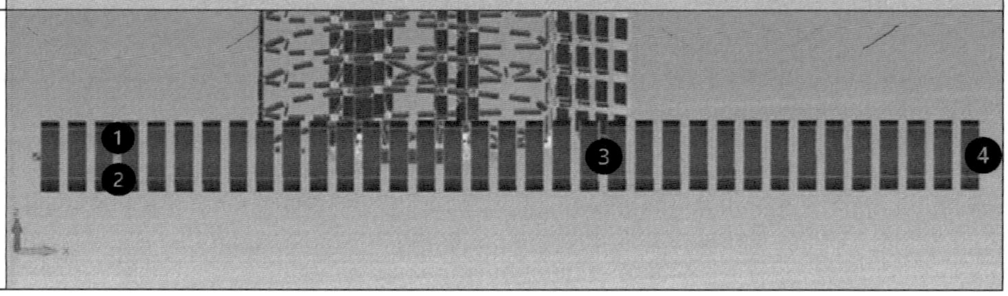

Table 2. Mass configurations of the wing-tip pod

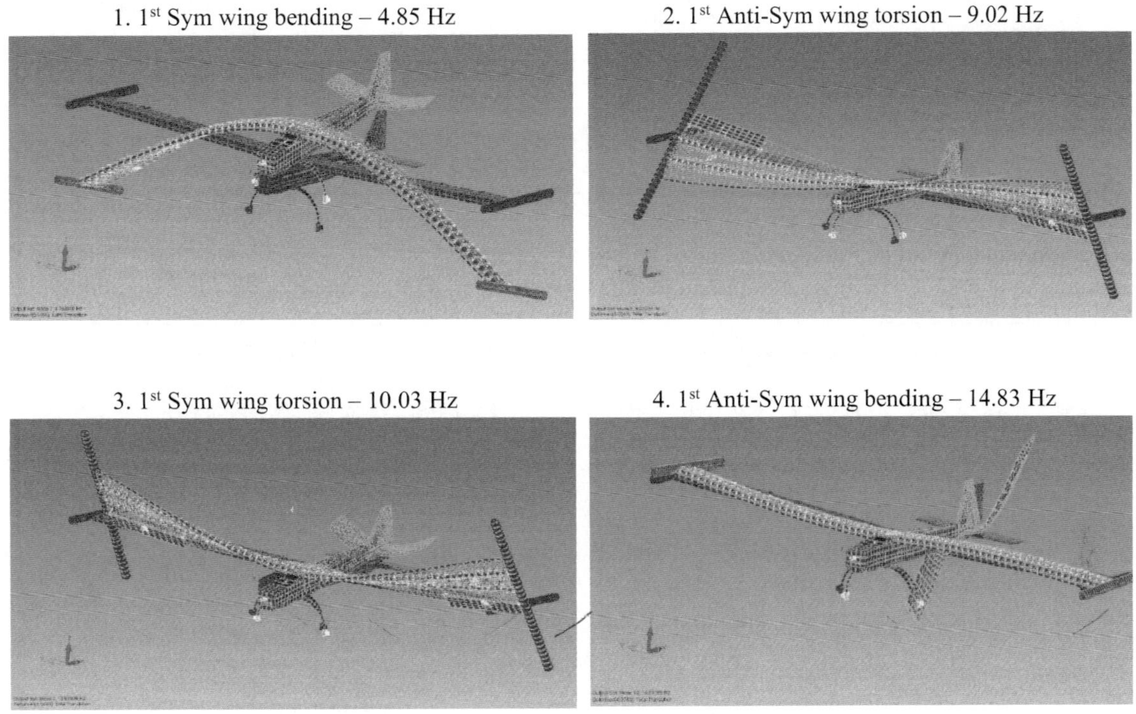

Figure 2. Natural modes and frequencies of the modified flight-test vehicle

IV. Aeroelastic Analysis and PFM Implementation

A. Classic Flutter Calculations

To capture unsteady aerodynamic forces, we use ZAERO to build panel meshes for the fuselage, wings, tail, and pods. The structural normal mode displacements from the FEMAP model are spline-interpolated to the aerodynamic mesh. ZAERO's g-method [4] scans velocity from 0 up to 70 m/s to calculate the variations of aeroelastic damping and frequencies. Figures 3–4 show the damping and frequency plots vs. velocity, and the anti-symmetric flutter mode shape at 44.16 m/s, where a damping branch crosses the zero line. The corresponding torsion-roll flutter frequency is 6.91 Hz. As summarized in Table 3, the first symmetric flutter arises near 62.42 m/s. With the modified configuration, the first antisymmetric flutter speed is 44.91 m/s, and the symmetric one is 66.92 m/s. The model changes placed the lowest nominal flutter velocity in the 40–45 m/s range, where flutter was identified during the second flight campaign.

Configuration	Mechanism Symmetry	Flutter velocity (m/s)	Flutter frequency (Hz.)	Flutter mechanism
Nominal	Anti	44.16	6.91	Anti-symmetric wing torsion-roll
	Sym	62.42	6.75	Symmetric wing torsion bending
Modified	Anti	44.91	6.73	Anti-symmetric wing torsion-roll
	Sym	66.92	6.17	Symmetric wing torsion bending

Table 3. Results summary for Nominal and Modified Configurations in ZAERO.

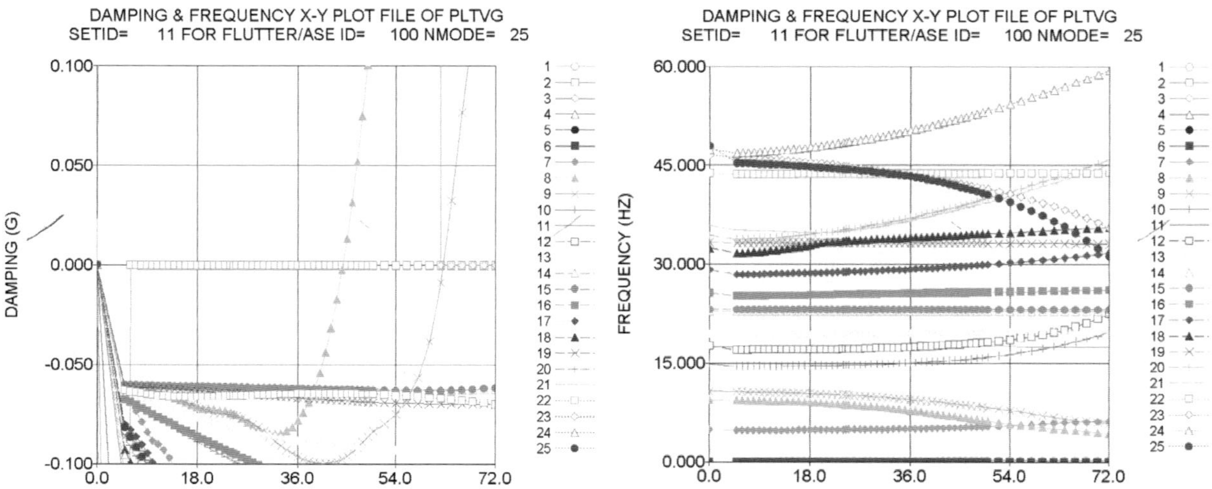

Figure 3. Nominal damping vs. velocity (Left) & Nominal frequency vs. velocity (Right) from ZAERO.

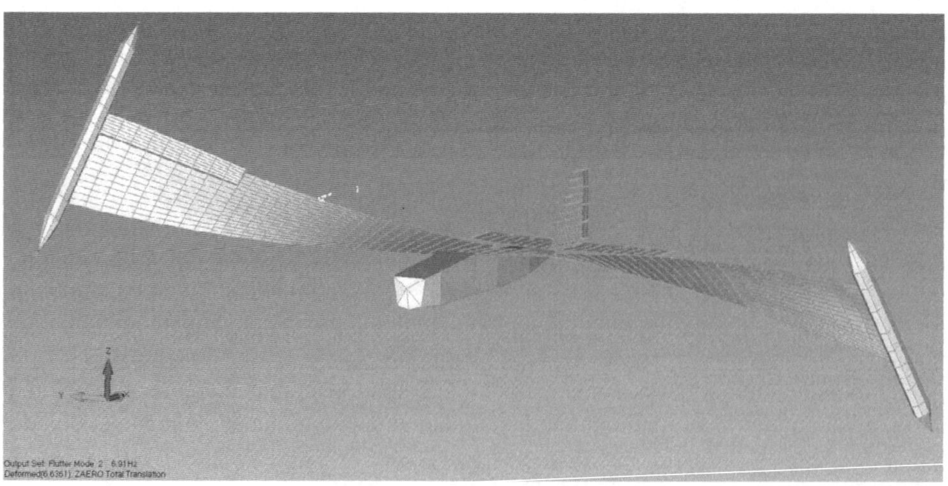

Figure 4. Nominal first flutter mode (44.16 m/s, 6.91 Hz) in the aerodynamic grid

B. DynResp PFM flutter analysis

A crucial phase of this work involved using the DynResp to predict flutter onset under forced excitation, then identifying how an added mass parameter (p_f) shifts the flutter boundary. While ZAERO provided classical flutter speeds via g-method eigenvalue solutions, DynResp allows us to replicate the *actual flight test procedure*: a known excitation force is applied (through the accelerated moving mass), and the system's frequency response is scanned at various airspeeds. By searching for the velocity at which the parametric mass effects are "removed", the true flutter velocity of the *nominal* UAV is found, all without physically crossing the stability boundary during the real flight. This section shows the results of the SISO-PFM approach in DynResp, illustrates the resulting Δp_f vs. velocity plots for each configuration, and clarifies why multiple shaker–sensor pairings were also examined before moving on to direct-force FRF analyses in the next section.

The SISO-PFM Setup started by adopting a single "right shaker–right sensor" pairing in the wing-tip pod, closely mirroring the historical SISO scheme from earlier experiments. In the FE model, that shaker location corresponds to a forward grid node (for example, Grid 1535) where the Parametric Flutter Parameter mass p_f is conceptually introduced. By specifying a small co-located force and measuring (in the model) the vertical acceleration at that same node, DynResp constructs frequency-by-frequency responses for each tested airspeed. The solver then uses the SISO formula of Eq. (2) to track how the system would move from stable to unstable if this mass or stabilizing parameter were factored out in post-processing.

The DynResp runs also use the same Nominal vs. Modified configurations as presented in Table 3, with the 0.04 and 0.06 structural damping coefficients. Table 4 compares the resulting DynResp flutter speeds against ZAERO. For the nominal UAV, the anti-symmetrical torsion–roll flutter emerges at 44.14 m/s, matching ZAERO's 44.16 m/s. A second symmetric flutter appears near 62 m/s, again within 1 % of ZAERO's value. Similar consistency appears in the modified case, with the first flutter near 44.83 via DynResp and 44.91 m/s via ZAERO and the second flutter at 66.7 and 66.9 m/s accordingly. Figure 5 shows the FRF of the corresponding DynResp runs of the modified configuration. The red x marks the flutter speed recognition, while the turquoise marks the zero-phase crossings and their corresponding gains. It may be noticed that the phase lines approach the zero-phase crossings at very small angles, which might yield sensitive results when performed with air-turbulence noise.

		ZAERO		**DynResp (SISO)**	
Case	**Symmetry**	Flutter velocity (m/s)	Flutter frequency (Hz.)	Flutter velocity (m/s)	Flutter frequency (Hz.)
Nominal	**Anti**	44.16	6.91	44.14	6.91
	Sym	62.42	6.75	62.05	6.84
Modified	**Anti**	44.91	6.73	44.82	6.74
	Sym	66.92	6.17	66.70	6.23

Table 4. Summary Table of Flutter Results of Nominal and Modified Configurations Between ZAERO and DynResp

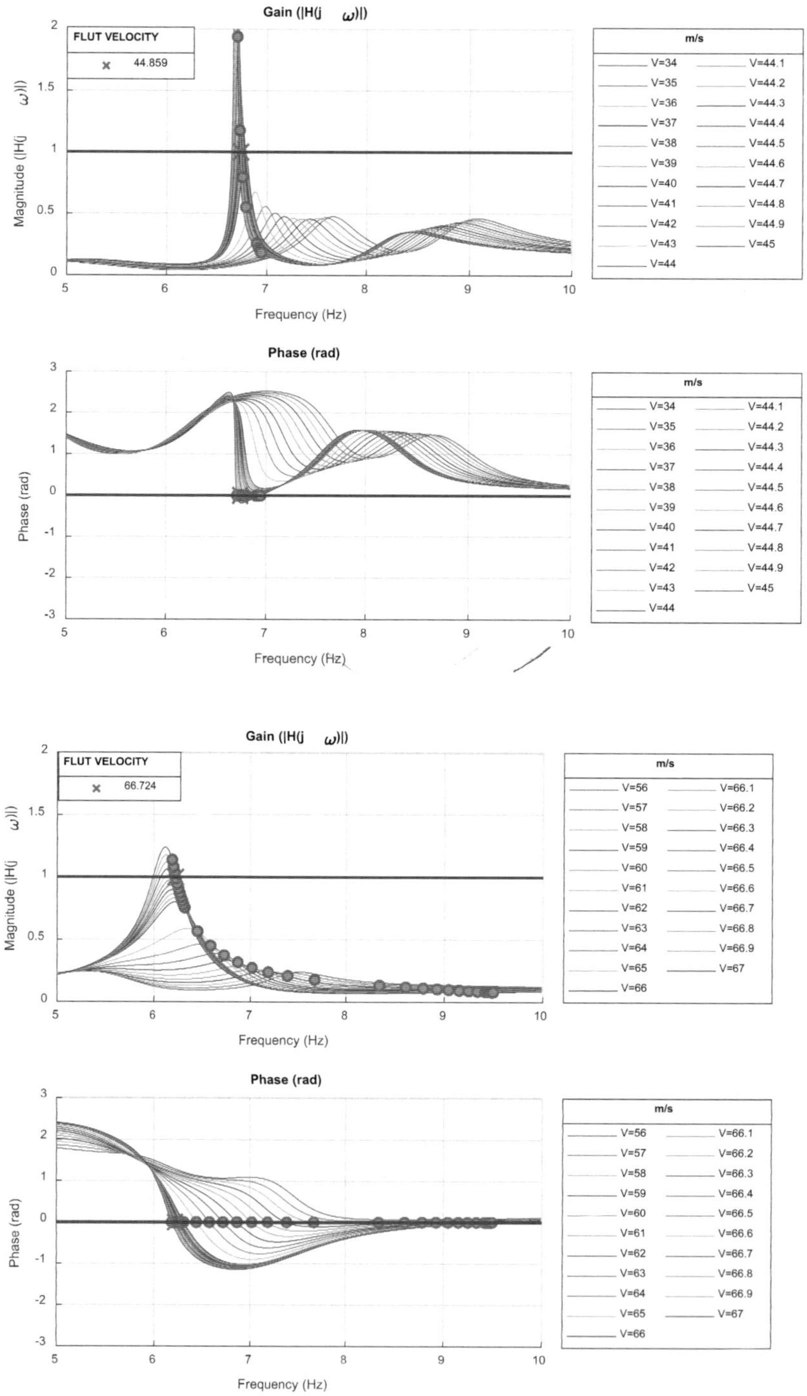

Figure 5. Response Bode plot of the modified configuration using PFM method.
(The above is the anti-symmetrical flutter response and below it the symmetrical response)

Parametric increments, Δp_f of Eq. (5), vs. velocity plots may be extracted to show how changing the stabilizing mass affects the flutter velocities. DynResp produces parametric loops that compute a Δp_f at the zero-phase crossover. Figure 6 illustrates Δp_f vs. velocity curve of the modified configuration, highlighting the lower flutter at 44.8 m/s and a higher symmetrical branch at 66.7 m/s, where the curve crosses $\Delta p_f = 0$. A negative slope near 44.8 m/s implies that removing just 50–100 g from the forward tip may increase the flutter speed by 1–2 m/s, an important design factor if the operator intends to remain below certain flight speeds. Conversely, the symmetric flutter near 66.7 m/s displays a somewhat less steep slope, suggesting that small mass variations have a large effect on that higher flutter branch.

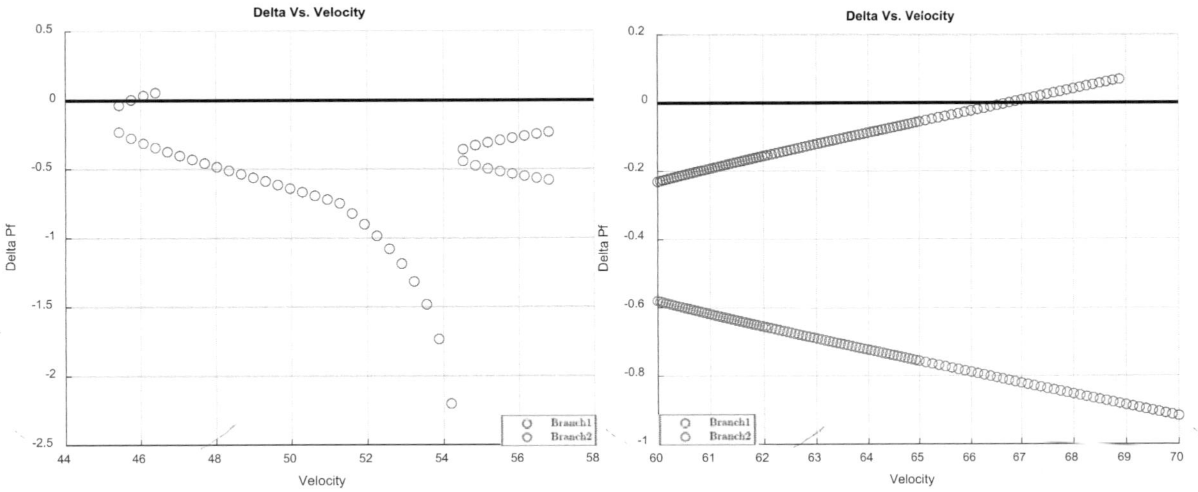

Figure 6. a Δp_f vs. velocity plot of Modified Flutter velocity calculation using DynResp

C. Multiple Shaker–Sensor Pairings

Although the single "right shaker–right sensor" scheme sufficed to confirm flutter speeds, we briefly examined other pairings (e.g. right shaker–left sensor, left shaker–right sensor, left shaker–left sensor) to simulate more realistic flight conditions with two pods and multiple accelerometers. In actual flights, the UAV is flying with both shakers and recording data in both wingtips. The shaking itself is also made asymmetrically in order to excite both symmetric and anti-symmetric responses simultaneously.

By running DynResp separately with each pairing (still SISO each time), we confirmed that the predicted first flutter velocity typically differs by less than 0.3 m/s across pairings, and the flutter frequency differs by less than 0.1 Hz. Hence, the UAV's fundamental flutter boundary around 44.14 m/s is robustly identified irrespective of which single channel is used. This consistency is significant: it mirrors how real flight data might combine left and right signals to isolate symmetrical or anti-symmetrical modes. Therefore, the multi-pair analysis helped validating that small sensor or shaker differences do not obscure the main conclusion about where flutter emerges.

Key Observations of the DynResp analysis: First, the DynResp SISO-PFM approach aligned within ~1 % of ZAERO for both nominal and modified setups, proving that forced-response parametric logic yields the same flutter boundary as an eigenvalue scan. Second, multi-pair analysis showed that sensor–shaker pairing does not drastically shift the flutter predictions. Third, for each configuration, we produced the Δp_f vs. velocity curves to see how the parametric mass removal triggers instability. These results pave the way for direct-force FRF analysis in the following section, where we do not rely on the SISO-PFM loop but instead measure responses to direct force applied at the shaker nodes, measured both right and left responses and combined them into symmetric and anti-symmetric SISO signals, enabling us to compare them to the flight-test data analysis.

V. Flight-Test FRF Results and Comparisons

A. Window/Overlap Sensitivity in Flight Data

OSU's flight test at 42.5 m/s, reached via a dive, gathered accelerometer signals on each wing-tip pod. These signals are processed with a transfer function estimator ("tfestimate") that depends on two key parameters: window size (e.g., 1200 or 2000 samples) and overlap (e.g., 800, 1100, 1600). Because the flutter "event" is mild, the amplitude of the flutter mode in the range of 6–9 Hz is so subtle, that different estimator parameters can either sharpen or smear effect and cause up to ~1 Hz shift in the zero-phase crossing and the corresponding gain peak. Figure 7 shows the anti-symmetrical FRF for multiple window–overlap pairs: W=1200, O=800 versus W=2000, O=800, W=2000, O=1600, etc. The magnitude and phase curves differ enough that one might conclude flutter emerges at 8.0 Hz in one pipeline or 7.2 Hz in another.

Hence, the borderline nature of near-flutter in real flight makes consistent data processing critical. Larger windows (2000) with heavier overlap (1100–1600) produce smoother, more stable FRFs but risk merging narrow peaks. Smaller windows highlight sharper spikes but include more random noise. In practice, in their work [5] OSU's standard is W=1200, O=800, but if post-processing is varied, the perceived flutter margin can shift. These observations underscore the caution needed when matching flight data with numeric results. In addition, this effect emphasizes the importance of lowering the flutter speed into the flight envelope of 30-40 m/s, in order to achieve longer and more robust flutter recordings at steady horizontal flight.

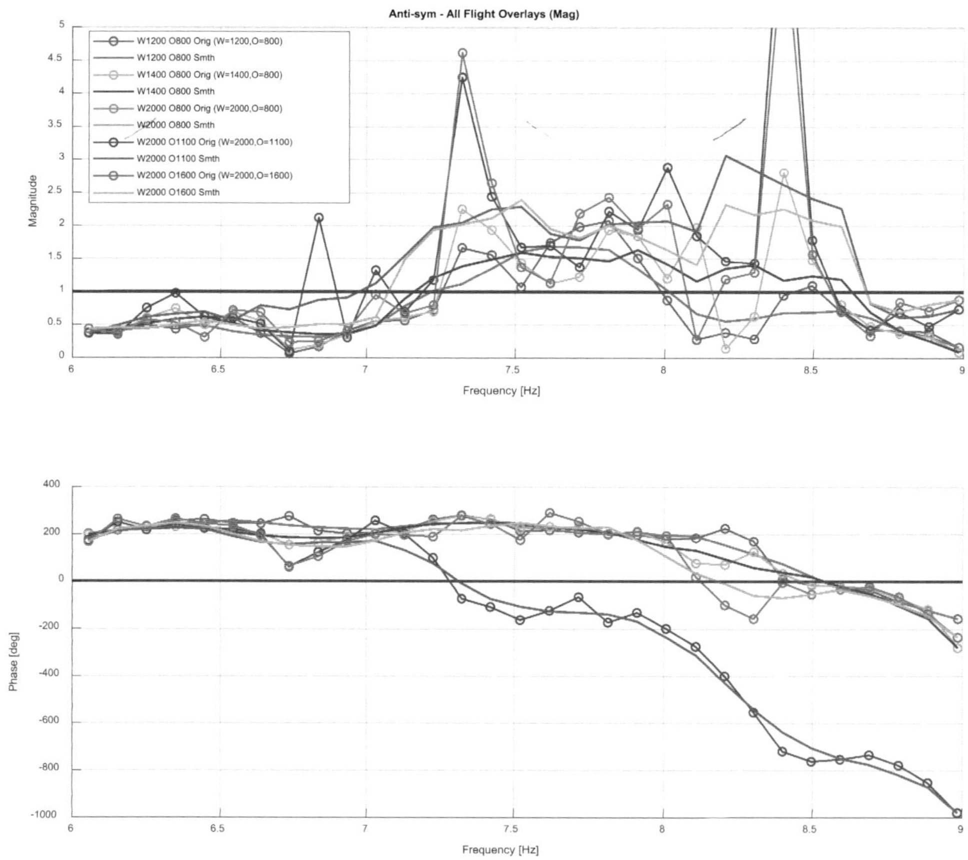

Figure 7. Comparison of anti-sym. FRFs with different window/overlap combos and unwrapped phase.

5.2 Direct-Force Excitation and Comparison with Flight Data at 42.5 m/s

In addition to the PFM-based analyses presented in earlier sections, the present work employs a direct-force approach in DynResp to replicate the forced responses measured in real flight tests. Instead of adding or removing a virtual stabilizing mass as in the PFM method, we specify literal force inputs at specific nodes, letting DynResp compute the system's steady-state oscillations over a predefined frequency range. This "non-PFM" route simplifies the interpretation of forced vibrations because the solver outputs direct frequency responses (as complex response) like what one observes when a physical shaker is activated at mid-flight. By comparing these purely numerical FRFs against FRFs processed from the measured signals of the UAV's second flight campaign, where a wing-tip shaker excited the structure while at 42.5 m/s dive. We can assess how well the final finite element model and flutter analysis captures the platform's actual dynamics.

The workflow for running a direct-force sweep in DynResp begins by specifying a range of flight velocities (for instance, from 34 m/s up to 45 m/s in increments of 0.1 m/s) and a frequency grid (we choose 6–9 Hz because we anticipate the flutter frequency to occure in this range). DynResp then applies sinusoidal inputs at each frequency step, solves the forced response, and writes a single *.f58 file containing all frequencies for each velocity. For each sensor node of interest, real and imaginary parts of the accelerations are stored. Because our aim is to overlay the computed FRFs with flight data, the solver's outputs need to be processed into magnitude and phase at each frequency. We do this by parsing the *.f58 file and converting Re and Im columns to amplitude and a phase angle. Once these data are collected for a sweep of speeds, we can produce Bode plots that reveal how the UAV's structure resonates or shifts toward instability as velocity increases.

In the actual flight campaigns, two electromagnetic shakers inside the UAV's wing-tip pods imparted excitations across 2–20 Hz [5]. The measured forces and accelerations, when transformed via transfer function estimator, yielded flight-test FRFs at discrete frequencies. To ensure our numeric approach mirrors this experimental arrangement, we applied a direct force at the pod node in the DynResp model (the same location used for the PFM approach but now set up to transmit a literal sinusoidal load rather than a stabilizing mass parameter). This yields a response in the wing and fuselage degrees of freedom, which we save as amplitude/phase across the velocity range. We then compare these numerical FRFs to flight logs at a specific speed, in this case, 42.5 m/s where the UAV was near-flutter in a dive. Because real flight logs can be sensitive to how one processes the raw time-series data, we use a window size of 2000 samples and an overlap of 1100 to achieve a smoother frequency spectrum, consistent with the flight team's more refined post-processing attempts.

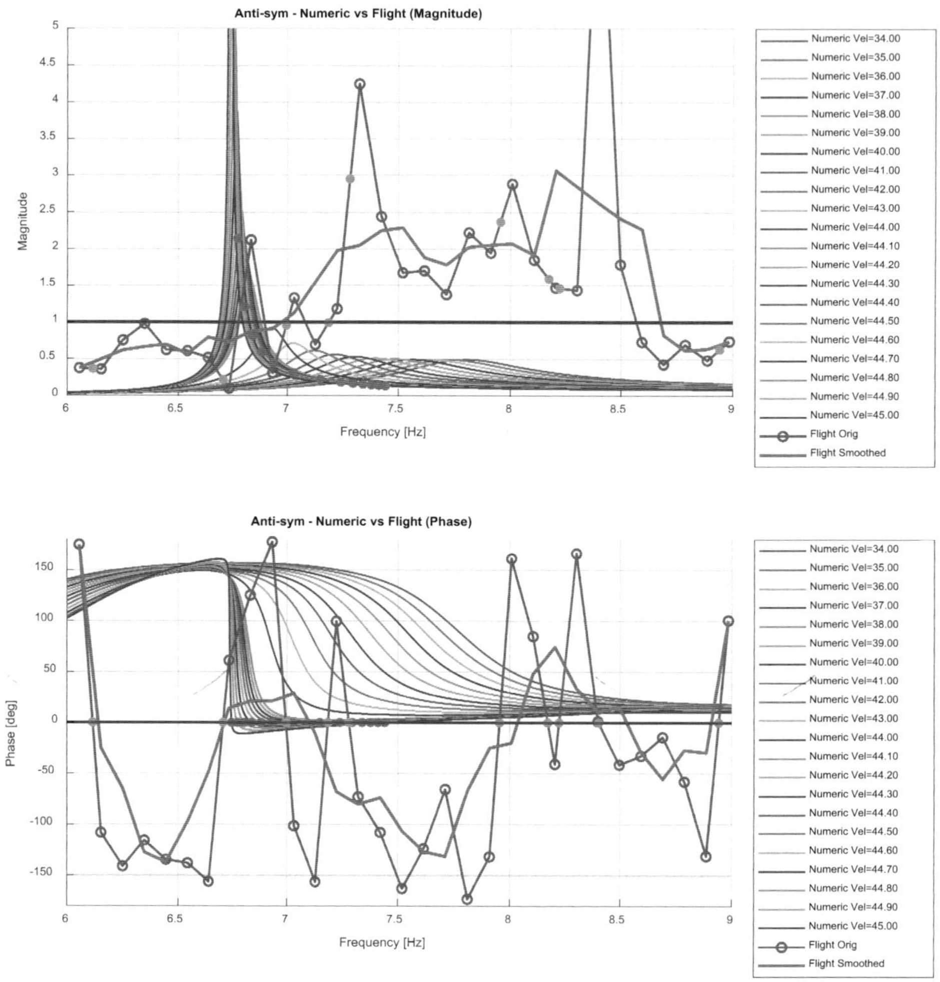

Figure 8. Anti-Symmetrical FRF Comparison of Flight Data Vs Numerical DynResp Analysis of the Modified Configuration (W2000 & O1100)

Figure 8 illustrates the result of this overlay. On the same axis, we plot both the measured flight FRF at 42.5 m/s, obtained by the OSU code with the chosen window/overlap, and multiple DynResp runs at speeds from 34 m/s to 45 m/s in small increments, focusing on the anti-symmetric channel because the flight data repeatedly indicated that torsional-roll coupling is the first flutter mechanism. The red markings are the numerical zero-phase crossovers and their corresponding gain. The green markings are the flight data zero-phase crossovers and their corresponding gains.

In many cases, the flight data FRF curve maintains a strong resonance near 6.8 Hz, 7.3 Hz and 8.4 Hz, whereas the numerical main peak is at 6.8 Hz. Although there are similar peaks, the major peaks differentiate at about 1.5 Hz. That difference suggests that while the updated finite element model, GVT calibrations, and direct-forcing approach capture the general trend and predict flutter across several industrial tools, some structural parameters, might be slightly under or overestimated, or another option is that the flutter event wasn't "dominant" enough to be identified with different transfer function estimator parameters. This discrepancy does not invalidate the approach, rather it pinpoints an area for further refinement in the model and points of focus in following test flights to be conducted.

Because the UAV never actually exceeded 44–45 m/s in steady flight, the PFM approach is able to validate that the nominal "no-shaker" configuration would become unstable around that range, while physically the system was tested in a safer scenario by retaining an effective forward mass. It is known that the flutter margin is shown to be at least 20 % then the UAV might satisfy certification guidelines without requiring direct full-flutter flights [6]. This synergy between advanced FE modeling, GVT correlation, direct-force response analysis, and carefully processed flight FRFs provides a robust method for UAV flutter clearance in a fast numerical way.

VI. Conclusions and Future Steps

The PFM method offers a powerful framework for investigating UAV flutter boundaries without risking uncontrolled divergence. By artificially stabilizing the aircraft with an added parameter p_f , commonly a forward tip-pod mass, the real flutter boundary is shifted to higher speeds. Test flights can therefore remain below the original critical speed, obtaining frequency-response data that is post-processed to remove the effects of p_f. This approach proved effective in capturing near-flutter phenomena at 42.5 m/s for a UAV that, in its purely nominal state, likely flutters around 44–45 m/s. The collaboration of GVT calibrations and flight data ensures the updated finite-element model, GVT-based modal damping coefficients, corrected torsional moduli, and an aft extension with 220 g mass, yields flutter predictions consistent with real events.

Despite certain challenges in flight data analysis, such as short time records, noise, or the need to dive for higher speeds the overlay comparisons with DynResp direct force solutions reveal that an anti-symmetrical torsion-roll mechanism dominates near 6.75 Hz. Minor differences in overlap/window of the flight data transfer function estimator can shift the measured peak up by 1 Hz, cautioning that "mild" flutter is highly sensitive to the data pipeline. Future research and works aims to completely re-design the wing structure further to lower flutter velocity so the UAV can reach flutter within its 30–40 m/s envelope while in stable level flight, thereby reducing the reliance on dives and capturing more robust flutter onset signals. Additional plans include analyzing alternative, more forward mass placements that might further shift flutter speed downward for both more robust flutter recording and simultaneously safer testing.

In summary, the combination of numerical modeling, parametric mass margins, and thoughtful flight testing stands as an effective way to easier and more efficient UAV aeroelastic clearance. By systematically bridging the gap between GVT results and real-flight FRFs, this methodology offers a strong foundation for both academic investigations and future certification processes, especially when a 20 % margin demonstration is required by regulators [6]. The UAV developer thus gains confidence in the structural design, ensuring that operational speeds remain well below the dangerous flutter boundary, or that the system can be stabilized if flutter does approach. The approach is applicable to a wide range of flexible unmanned or manned aircraft needing rapid or cost-effective flutter clearance.

References

[1] Heidersbach, R., Seth, D., McCrink, M., Karpel, M., "Safe Flutter Flight-Testing of an Unmanned Aerial Vehicle Utilizing Parametric Flutter Margin," *AIAA SCITECH 2023 Forum*, Washington D.C., January 2023. DOI: 10.2514/6.2023-2071

[2] Karpel, M., McCrink, M., Heidersbach, R., and Seth, D., "Safe Wind Tunnel Flutter Tests Using a Shaker and Mass Margin," *International Forum on Aeroelasticity and Structural Dynamics (IFASD)*, Madrid, June 2022.

[3] Roizner, F., Raveh, D. E., and Karpel, M., "Safe Flutter Tests Using Parametric Flutter Margins," *Journal of Aircraft*, Vol. 56, No. 1, 2019, pp. 228–238, doi: 10.2514/1.C035045.

[4] Chen, P. C., "Damping Perturbation Method for Flutter Solution: The g-Method," *AIAA Journal*, Vol. 38, No. 9, 2000, pp. 1519–1524.

[5] Heidersbach, R., et al., "Safe Flutter Flight Test Using Parametric and Predictive Flutter-Margin Methods," *Proceedings of the AIAA SciTech 2024 Forum*, January 2024.

[6] Roizner, F., and Karpel, M., "Parametric Flutter Margin Method for Aeroservoelastic Stability Analysis," *AIAA Journal*, Vol. 56, No. 3, 2018, pp. 1011–1022.

ADVANCEMENTS IN IN-FLIGHT AEROELASTIC SENSING: REAL-TIME PREDICTION OF DEFLECTION, LOAD, AND RIGID ANGLE OF ATTACK USING MACHINE LEARNING FROM GROUND TEST DATA

Ido Hauzer*, Daniella E. Raveh

Technion – Israel Institute of Technology,
Technion City, Haifa, 3200003
Israel

ABSTRACT

The pursuit of lighter aircraft has driven a shift towards more flexible structures, leading to in-flight wing deformations that significantly affect aerodynamic characteristics, load distribution, and dynamic responses. Precisely sensing these deformations during flight tests is vital for establishing and verifying the design load envelope, validating analysis tools like flow and structural simulations, and exploring new control schemes. This study presents a Machine-Learning (ML) approach to map strain measurements into displacement and load distributions in highly flexible structures. Additionally, these strain data, combined with an aerodynamic model, are used to estimate the flight's rigid angle of attack (AoA).

In the proposed study, a Neural-Network (NN) model is trained to predict a wing's shape ([1], [2], [3], [4], [5], [6], [7], [8]) and load distribution ([9], [10], [11], [12], [13]) under varying distributed loading. The workflow comprises two stages. In the calibration stage, ground tests are conducted where responses to concentrated loads are recorded and used for model training and testing. In the application stage, the wing functions as a measuring device, where strains that develop under distributed loads are used to estimate deformations and loads in bench testing, wind tunnel experiments, or flight testing. The data can also be used to construct a detailed aerodynamic load distribution, and, when the latter is compared to an existing aerodynamic load distribution model, the real-time rigid AoA [13] can be estimated.

The application of the proposed methodology is demonstrated with the Pazy wing model [14], a benchmark, very flexible wing that can undergo 50% span deformation under aerodynamic loads. The NN model is first created using numerical data from Finite-Element Analysis, simulating a ground test scenario, and is subsequently validated through wind tunnel tests. The model's predictive accuracy is rigorously scrutinized across a wide range of load scenarios. For example, Figure 1 shows wind tunnel measured and predicted bending moment along the span, the constructed aerodynamic model, and the measured (by a load balance) and predicted moment values at the wing root. Figure 2 compares the estimated and measured rigid AoA during an AoA sweep wind tunnel test. The full paper will detail the methodology used to construct these predictions and further explore the models' capabilities.

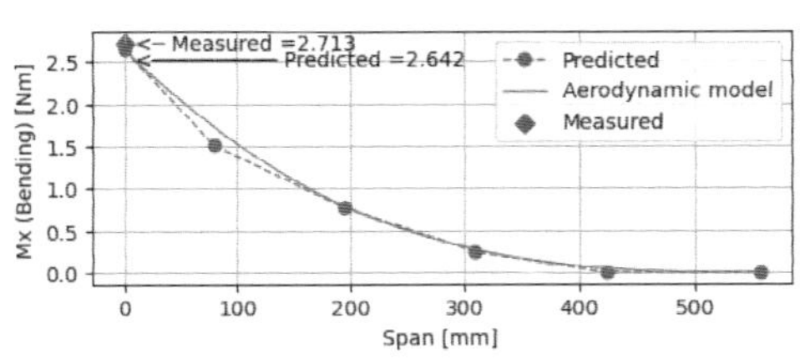

Figure 1 - bending moment along span

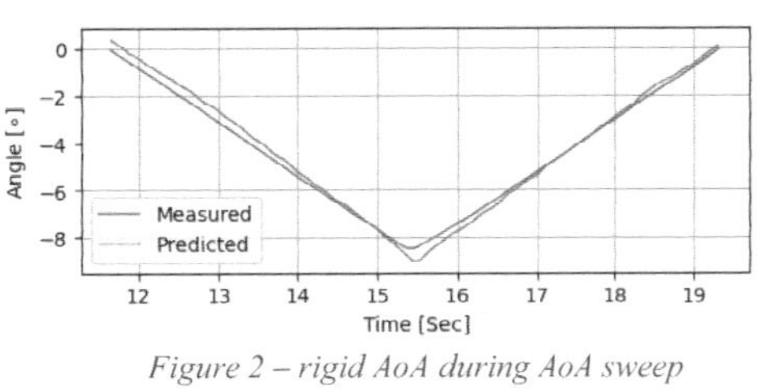

Figure 2 – rigid AoA during AoA sweep

Manuscript 41 submitted to 64th Israel Annual Conference on Aerospace Sciences (IACAS). Received January 5, 2025.

[1] A. Drachinsky and D. E. Raveh, “Modal Rotations: A Modal-Based Method for Large Structural Deformations of Slender Bodies,” *AIAA Journal*, vol. 58, no. 7, pp. 3159–3173, Jul. 2020, doi: 10.2514/1.J058899.

[2] A. Skafte, M.L. Aenlle and R. Brincker, “A General Procedure for Estimating Dynamic Displacements Using Strain Measurements and Operational Modal Analysis,” *Smart Materials and Structures,* vol. *25, no. 2*, p. 025020, Jan 2016.

[3] W.L. Ko, W.L. Richards, and V.T. Fleischer, “Applications of KO Displacement Theory to the Deformed Shape Predictions of the Doubly-Tapered Ikhana Wing.” No. NASA/TP- 2009-214652. Oct. 2009.

[4] C.G. Pak, “Wing Shape Sensing from Measured Strain,” *AIAA Journal*, vol. 54, no. 3 ,pp. 1068-1077, Mar. 2016.

[5] C.G. Pak, “Linear and Geometrically Nonlinear Structural Shape Sensing from Strain Data,” *AIAA Journal,* vol. 61, no. *2*, pp. 907–922, Nov. 2022, doi: 10.2514/1.J062165.

[6] M. Gherlone, C. Priscilla, and M. Mattone, “Shape Sensing Methods: Review and Experimental Comparison on A Wing-Shaped Plate,” *Progress in Aerospace Science,* vol. 99, pp. 14-26, May 2018, doi: https://doi.org/10.1016/j.paerosci.2018.04.001.

[7] S. Sefati, C. Gao, I. Iordachita, R. H. Taylor, and M. Armand, “Data-Driven Shape Sensing of a Surgical Continuum Manipulator Using an Uncalibrated Fiber Bragg Grating Sensor,” *IEEE Sensors Journal,* vol. 21, no. 3, pp. 3066-3076, Oct. 2020, doi: 10.1109/JSEN.2020.3028208.

[8] T. L. T. Lun, K. Wang, J. D. L. Ho, K.H. Lee, K. Y. Sze, and K.W. Kwok, “Real-Time Surface Shape Sensing for Soft and Flexible Structures Using Fiber Bragg Gratings,” *IEEE Robotics and Automation Letters*, vol. 4, no. 2, pp. 1454–1461, Apr. 2019, doi: 10.1109/LRA.2019.2893036.

[9] C. W. Coates and P. Thamburaj, “Inverse Method Using Finite Strain Measurements to Determine Flight Load Distribution Functions,” *Journal of Aircraft*, vol. 45, no. 2, pp. 366–370, Mar. 2008, doi: 10.2514/1.21905.

[10] P. M. Trivailo and C. L. Carn, “The Inverse Determination of Aerodynamic Loading from Structural Response Data Using Neural Networks,” *Inverse Problems in Science and Engineering*, vol. 14, no. 4, pp. 379–395, Jun. 2006, doi: 10.1080/17415970600573692.

[11] T. Nakamura, H. Igawa, and A. Kanda, “Inverse Identification of Continuously Distributed Loads Using Strain Data,” *Aerospace Science and Technology*, vol. 23, no. 1, pp. 75–84, Dec. 2012, doi: 10.1016/j.ast.2011.06.012.

[12] D. Wada, Y. Sugimoto, H. Murayama, H. Igawa, and T. Nakamura, “Investigation of Inverse Analysis and Neural Network Approaches for Identifying Distributed Load using Distributed Strains,” *Transactions of the Japan Society for Aeronautical and Space Sciences*, vol. 62, no. 3, pp. 151–161, 2019, doi: 10.2322/tjsass.62.151.

[13] D. Wada and M. Tamayama, “Wing Load and Angle of Attack Identification by Integrating Optical Fiber Sensing and Neural Network Approach in Wind Tunnel Test,” *Applied Science*, vol. 9, no. 7, pp. 1461, Apr. 2019, doi: 10.3390/app9071461.

[14] O. Avin, D. E. Raveh, A. Drachinsky, Y. Ben-Shmuel, and M. Tur, “Experimental Benchmark of a Very Flexible Wing”, *AIAA Journal*, vol 60, no.3, pp. 1745-1768, Mar. 2021. doi: 10.2514/1.J060621.

Network of symmetric periodic families in the Hill 3-body problem

Alexander Batkhin
Department of Aerospace Engineering,
Technion – Israel Institute of Technology,
Haifa, Israel
Email: batkhin@technion.ac.il

Abstract

A technique of Conley–Zehnder indices is applied for investigation of interconnections of the basic families of periodic orbits with maximal numbers of symmetries of the well-known Hill problem. These basic families are g, f – families of planar direct and retrograde periodic orbits, and $\mathcal{B}_0$ – family of rectilinear vertical consecutive collision orbits. The relations among families of periodic orbits are provided by families of spatial symmetric periodic orbits, which makes k-covering at the bifurcation points. All families form a common network and can be represented as well-organized bifurcation graphs of the interconnectedness.

The Hill three-body problem (Hill3BP), a limiting case of the circular restricted three-body problem (RTBP), is a well-known model which provides an approximation of the dynamics of the infinitesimal body in the vicinity of the smaller primary. In its original application, George Hill reformulated the lunar theory and discovered a periodic solution with a period equal to the synodic month of the Moon. There are a lot of applications of Hill's approach such as capturing in the dynamics of natural or artificial satellites, distant moons of asteroids, low-energy escaping trajectories, and frozen orbits around planetary satellites. Hill3BP's periodic solution can be continued to RTBP or even into three-body problem solutions and thus can be used in astrodynamical projects.

Hill3BP's equations of motion are invariant under discrete group of (anti-)symplectic symmetries of the extended phase space. So, all solutions to Hill3BP can be divided into groups with different numbers of symmetries. The symmetry of a periodic solution plays an essential role, since it allows one to investigate it numerically for only parts of the period.

There are 3 families, called *basic families*, whose orbits are simple and have the largest number of symmetries: g and f are families of planar direct and retrograde satellite orbits and $\mathcal{B}_0$ is a family of vertical collision orbits. Other important families are Lyapunov families a and c emanating from the libration points L_1 and L_2 and family g' appeared after symmetry-breaking bifurcation of the family g. Earlier was shown [1] that all these families are connected to each other by families of spatial periodic orbits and form a kind of common network. Current work significantly extends these results by systematically applying the technique of Conley–Zehnder indices.

The summary of the current investigation [2] is shown in the table, where each row the integer n indicates each n-th cover bifurcation of the underlying family in the first row.

g	g'	$\mathcal{B}_0^{\pm}$	f	f_3	halo
	1				2
1		2			
2	2	3			
3	3	4	5	1	
4	4	5	6	2	
5		6	7		

As a consequence, the following connections were discovered: at bifurcation points between n-th cover of the families g, $n+1$-th cover of $\mathcal{B}_0^{\pm}$ and $n+2$-th cover of f, for $n = 3, 4, 5$. More detailed investigations of out-of-plane bifurcations demonstrate a complex structure of connections between the family g and a set of families of planar periodic solutions obtained from the family f with different numbers of covering.

References

[1] Batkhin A.B. and Batkhina N.V. Hierarchy of periodic solutions families of spatial Hill's problem // *Solar System Research.* 2009. **43**(2), 178–183. DOI:10.1134/s0038094609020105.

[2] C. Aydin and A. Batkhin, Studying network of symmetric periodic orbit families of the Hill problem via symplectic invariants // arXiv:2410.21245, 2024, 57 p. DOI:abs/2410.21245.

Manuscript 62 submitted to 64th Israel Annual Conference on
Aerospace Sciences (IACAS). Received January 9, 2025.

Uncertainty Propagation Using Riemannian Geometry

Vladimir Martinusi*
Department of Aerospace Engineering
Technion – Israel Institute of Technology

Abstract

Errors in measurements and inaccuracies in models can disturb the initial conditions used in orbit simulations, resulting in trajectories that deviate from the actual orbit. This raises an important question: how can we effectively quantify and comprehend how these uncertainties evolve over time? This study applies the principles of Riemannian geometry to analyze the propagation of errors in orbital mechanics. By conceptualizing the state space as a Riemannian manifold, the research investigates how factors like geodesics and curvature influence the dispersion of uncertainties. This innovative approach offers a deeper understanding of error growth mechanisms and contributes to improving orbit prediction and navigation methods.

1 Presentation Overview

This research centers on **conservative dynamical systems**, which are free from external influences such as atmospheric drag or solar radiation pressure. Such constraints ensure the conservation of the system's total energy.

The motion is described using the Lagrangian:

$$\mathcal{L} = \frac{1}{2}\dot{\mathbf{q}} \cdot \mathbf{G}\dot{\mathbf{q}} - \mathcal{V}(\mathbf{q}), \tag{1}$$

where:

- $\mathbf{q}$: Generalized coordinates,
- $\dot{\mathbf{q}}$: Generalized velocities,
- $\mathcal{V}(\mathbf{q})$: Potential energy, dependent only on $\mathbf{q}$, in alignment with the conservative framework.

*Senior Visiting Lecturer. Email: vmartinusi@technion.ac.il

Manuscript 89 submitted to 64th Israel Annual Conference on Aerospace Sciences (IACAS). Received January 15, 2025.

By employing the **Legendre transform**, the state space is reformulated from $(\mathbf{q}, \dot{\mathbf{q}})$ to $(\mathbf{q}, \mathbf{p})$, where $\mathbf{p}$ represents the generalized momenta:

$$\mathbf{p} = \mathbf{G}\dot{\mathbf{q}}. \tag{2}$$

The Hamiltonian is then expressed as:

$$\mathcal{H} = \mathbf{p} \cdot \dot{\mathbf{q}} - \mathcal{L} = \frac{1}{2}\mathbf{p} \cdot \mathbf{G}^{-1}\mathbf{p} + \mathcal{V}(\mathbf{q}). \tag{3}$$

Using the scalar function:

$$s(\mathbf{q}) = \frac{1}{\mathcal{H} - \mathcal{V}(\mathbf{q})}, \tag{4}$$

and substituting $s\mathbf{G}^{-1} \to \mathbf{G}$, the geodesic equation in vector form is derived:

$$\frac{1}{2}\mathbf{p} \cdot \mathbf{G}\mathbf{p} = 1. \tag{5}$$

The metric tensor $\mathbf{G}$ contains the necessary information to determine geodesic deviation, influenced by the constant Hamiltonian $\mathcal{H}$, which represents the specific energy along the trajectory.

1.1 Key Definitions

Metric Tensor: The metric tensor $\mathbf{G}$ is a key element in Riemannian geometry, defining distances and angles on the manifold. In this study, it is used to measure the separation between nearby trajectories in the state space.

Geodesic Deviation: Geodesic deviation describes the divergence between two infinitesimally close geodesics, illustrating how initial errors in the state space propagate due to the curvature of the manifold.

2 Geometric Insights

The study derives important geometric quantities:

- **Riemann curvature tensor**, which characterizes the intrinsic curvature of the manifold,
- **Ricci tensor**, a contraction of the Riemann tensor that describes volume distortion,
- **Ricci scalar**, a single value summarizing the average curvature at a specific point.

These mathematical constructs provide a detailed characterization of the manifold's local curvature, offering critical insights into how uncertainties evolve within conservative dynamical systems.

Manuscript 89 submitted to 64th Israel Annual Conference on Aerospace Sciences (IACAS). Received January 15, 2025.

3 Practical Implications

Explicit derivations of geodesic deviations demonstrate the ways in which initial-condition errors spread, showing their direct implications for orbit prediction. By utilizing curvature information, this methodology paves the way for enhanced trajectory estimation and navigation accuracy.

4 Conclusions and Future Directions

This research establishes a connection between the abstract principles of Riemannian geometry and practical challenges in orbital mechanics. Potential future directions include:

- Extending the framework to non-conservative systems (e.g., those influenced by drag or radiation pressure),
- Conducting numerical simulations to validate theoretical results,
- Broadening the application scope to other domains, such as robotics and control systems.

Manuscript 89 submitted to 64th Israel Annual Conference on Aerospace Sciences (IACAS). Received January 15, 2025.

Constant-Magnitude Low-Thrust Orbital Transfer with Final-Approach Targeting

Michal Pushkov, Shribharath Balakrishnan, and Pini Gurfil

Department of Aerospace Engineering
Technion – Israel Institute of Technology, Haifa 3200003, Israel

Abstract

Precise satellite orbit control is crucial for numerous space missions, requiring both fuel efficiency and accuracy in the presence of orbital perturbations. Thus far, however, little attention has been given to developing low-thrust closed-loop orbit control with guaranteed stability, accuracy, and performance, while accounting for the constant-magnitude thrust constraint typical to electric propulsion (EP) systems.

This research addresses these challenges by developing a closed-loop controller for satellite orbit transfers based on the Milankovitch vectorial elements. These elements, comprised of the eccentricity and angular momentum vectors, constitute a non-singular set of orbital elements. The current study examines variable-thrust and constant-magnitude thrust configurations, with emphasis on the latter due to its applicability in EP systems. Whereas EP systems are highly fuel-efficient, their low constant thrust presents unique challenges, such as potential chattering. To address these issues, the proposed orbit control methodology integrates Lyapunov-based control, Linear Quadratic Regulation (LQR), and Sliding Mode Control (SMC).

Initially, a Lyapunov-based controller based on Milankovitch vectorial elements was designed for variable-thrust maneuvers, with asymptotic stability established using LaSalle's invariance principle. This result is then extended to constant-magnitude thrust systems using two discretization strategies which guaranteed convergence for planar transfers.

Each control approach operates within specific state-vector frameworks: Milankovitch elements for the Lyapunov-based controller, and relative position and velocity for LQR and SMC. An algorithm is developed to transition between these state vectors. Comprehensive simulations validate the proposed methods under orbital perturbations, including Earth's gravitational potential, lunisolar gravitational forces, solar radiation pressure, and atmospheric drag. Two scenarios were analyzed: low Earth orbit (LEO) to geostationary Earth orbit (GEO) transfers, and low Earth orbit-raising. The performance was evaluated by considering maneuver time, fuel consumption, and accuracy.

In the LEO-to-GEO transfer scenario, several effective solutions were identified, balancing among accuracy, transfer time and ΔV. In low Earth orbit-raising, the Lyapunov controller achieves near-optimal performance.

Manuscript 43 submitted to 64th Israel Annual Conference on Aerospace Sciences (IACAS). Received January 6, 2025.

Optimal Control for a Stationary Solar Sail Around L1'

Naama Gilat*, Oded Golan†, and Vitaly Shaferman‡

Technion - Israel Institute of Technology, Haifa 3200003, Israel

January 9, 2025

Abstract

Climate change, a phenomenon that has disastrous effects on our planet, has prompted considerable research in the scientific community. Despite these efforts, a clear solution remains elusive. One potential solution under consideration is a space sunshade, placed at the first Sun-Earth Lagrange point (L_1). The sunshade would consist of a swarm of smaller spacecraft, each equipped with a reflective shielding surface that would divert the sun's radiation from Earth. This surface also serves as a solar sail propulsion system, which eliminates the need for propellant and allows very long operation time. The advantages of the swarm configuration are enhanced maneuverability, simplified operation, and is naturally redundant. The use of solar sails shifts the position of L_1 sunward (designated as L_1') and, by tilting the sail, allows the creation of an infinite number of new equilibrium points in the vicinity of the collinear equilibrium point L_1'. This work is part of the "Cool Earth" project at the Asher Space Research Institute. In this research, we have identified a manifold of equilibrium points for a given sail area and reflectivity, where each point on this surface is found for a different combination of sail angles. Subsequently, the sail's acceleration remains constant, and the system is underactuated. This work differs from previous studies by maintaining a fixed position at an equilibrium point, rather than orbiting around it (following a Halo or Lissajous orbit). After identifying the manifold, an optimal controller for a single spacecraft was constructed using an infinite-horizon linear quadratic regulator (LQR) based on model linearization around the nominal position on the manifold. This controller was evaluated in a simulation and was shown to be able to stabilize the spacecraft at any point on this surface when starting with small initial errors in position, velocity, and sail angles, confirming that this configuration provides sufficient control for station-keeping. We aim to extend this work by incorporating disturbances into the model, such as errors in the estimated position and velocity of the spacecraft and physical effects such as Moon gravity perturbation, time-varying solar radiation, and Solar Wind storms. This will allow us to evaluate how these factors influence the station-keeping capability of a single spacecraft and the robustness of the controller.

*Undergraduate student, Faculty of Aerospace Engineering. E-mail: naamagilat@campus.technion.ac.il
†Full Research Fellow, Faculty of Aerospace Engineering. E-mail: oded.golan@technion.ac.il
‡Associate Professor, Faculty of Aerospace Engineering. E-mail: vitalysh@technion.ac.il

Manuscript 63 submitted to 64th Israel Annual Conference on Aerospace Sciences (IACAS). Received January 9, 2025.

Technical Requirements for Optical Sensors in Autonomous Space Situational Awareness

Yuval Hammer* , Ido Ben Harosh† , Ivan Goncharov‡ ,
Ira Wolfson§ , and Elad Denenberg¶
Braude Academic College of Engineering, Karmiel, 2161002, Israel

The rapid growth of satellite deployments in Low Earth Orbit (LEO) necessitates autonomous collision avoidance capabilities for spacecraft. This paper investigates the technical requirements for optical sensors in autonomous Space Situational Awareness (SSA) applications, focusing on CubeSat implementations. The desired capability is detecting $1\ cm$ objects at distances of $500\ km$, but optical analysis reveals this would require currently unattainable $10\ nm$ pixel sizes. Given these technological constraints, we examined reduced detection distances by analyzing six representative satellite pairs using the SGP4 propagator at ranges of $500\ km$, $200\ km$, and $100\ km$. Our findings demonstrate that reducing detection distance from $500\ km$ to $200\ km$ decreases detection events by $70-80\%$ and observation durations from approximately $2\ s$ $0.5\ s$, with further reductions at $100\ km$. The results establish that while current technology cannot achieve optimal detection ranges, viable collision avoidance may be possible by focusing on larger objects ($\leq 10\ cm$) at reduced detection distances. However, this significantly impacts the frequency and duration of detection opportunities. These findings provide crucial parameters for implementing autonomous collision avoidance systems within current technological constraints.

I Introduction

The rapid growth of satellite deployments in LEO presents new challenges for spacecraft control. As mega-constellations like Starlink and OneWeb continue their deployment phases, the number of active satellites is expected to increase by an order of magnitude within the decade. This dramatic rise in orbital population increases collision risks and places additional demands on ground control networks. While ground stations have historically provided effective command and control for space missions, the increasing need for continuous monitoring and collision avoidance maneuvers suggests exploring complementary approaches. We propose developing autonomous spacecraft capable of independent navigation and collision avoidance decisions, reducing the operational load on ground station networks while maintaining space traffic safety.

*Student, Electrical and Electronic Engineering, Y.H@e.braude.ac.il
†Student, Electrical and Electronic Engineering, ido.ben.harosh@e.braude.ac.il
‡Student, Mechanical Engineering, ivan.goncharov@e.braude.ac.il
§Lecturer, Electrical and Electronic Engineering, wolfsoni@braude.ac.il
¶Lecturer, Mechanical Engineering, EladD@braude.ac.il.

This paper focuses on establishing the technical requirements for optical sensors in autonomous spacecraft. We investigate the necessary detection ranges, sensor dimensions, and operational concepts to enable successful collision avoidance maneuvers. The analysis considers various orbital scenarios to determine minimum detection distances and corresponding sensor specifications, providing a foundation for future autonomous SSA implementations.

We first examine existing detection technologies to understand the sensing requirements for autonomous spacecraft. Ground-based tracking currently employs radar systems for precise range measurements, optical telescopes for high angular resolution tracking, phased array systems for rapid orbital scanning, and electro-optical systems that combine optical and electronic capabilities [1–3]. While these systems effectively gather orbital data, relying on ground-based measurements and subsequent communication to spacecraft is incompatible with the goal of autonomous operation.

Space-based Light Detection and Ranging (LIDAR) systems have been suggested for orbital object detection [4,5]. However, LIDAR systems require significant power for laser operation, making them impractical for CubeSat implementations where energy resources are constrained.

Physical collectors, such as DRAGONS [6] or DEBIE [7], have provided valuable data on orbital debris through impact detection. However, these passive systems are unsuitable for collision avoidance applications, as they cannot detect potential threats before impact.

Space-based phased array radar systems [8, 9] use electronically steered beams for object detection and tracking. While effective for large spacecraft, these systems require substantial antenna arrays and power resources that exceed typical CubeSat constraints.

Optical sensing capabilities for space objects have been demonstrated in several studies. Flohrer [10] investigated space-based optical detection from LEO but focused on observing Geostationary Orbit (GEO) objects rather than the satellite's immediate environment. Spiller [11] demonstrated object detection using satellite star trackers. However, detection ranges for objects tens of centimeters in size were limited to tens of kilometers. Kruger [12] showed formation-keeping cameras could identify space debris, but this work centered on detecting nearby cooperative objects. While these studies establish the feasibility of space-based optical detection, they don't address the requirements for autonomous collision avoidance in LEO.

Implementing an optical sensor for autonomous spacecraft requires careful consideration of multiple design parameters: field of view specifications, environmental scanning methodology, detection range requirements, observation duration, sensor dimensions, and standard space-qualified system requirements. This paper examines these design decisions to establish technical requirements for optical sensors in autonomous SSA applications.

The remainder of this paper is organized as follows: Section II provides the theoretical background, examining both astrodynamical requirements and optical limitations. Section III presents our methodology for evaluating detection capabilities at different ranges. Section IV presents detailed results from analyzing six representative satellite pairs at various detection distances. Section V discusses the selected detection requirement parameters and future work.

II Background

II.A Optimal Astrodynamical Requirements

The motion of a satellite in Earth's gravitational field follows Newtonian mechanics. For a satellite, its position vector $\mathbf{r}$ relative to Earth's center follows the differential equation:

$$\ddot{\mathbf{r}} = -\frac{\mu}{r^3}\mathbf{r} \tag{1}$$

where $\mu = GM_E = 398600.43\ {}^{km^3}/_{s^2}$ is Earth's standard gravitational parameter and $r = \|\mathbf{r}\|$ is the magnitude of the position vector. The satellite's state is described by its position vector $\mathbf{r}$ and velocity vector $\mathbf{v} = \dot{\mathbf{r}}$.

For two objects in orbit, denoted by subscripts 1 and 2, their relative position vector $\mathbf{r}_{rel}$ is:

$$\mathbf{r}_{rel} = \mathbf{r}_2 - \mathbf{r}_1 \tag{2}$$

The relative velocity vector $\mathbf{v}_{rel}$ is:

$$\mathbf{v}_{rel} = \mathbf{v}_2 - \mathbf{v}_1 \tag{3}$$

For satellites in circular orbits, the orbital velocity is:

$$v = \sqrt{\frac{\mu}{r}} \tag{4}$$

For LEO satellites at an altitude of 500 km (orbital radius of 6878 km), the orbital velocity is 7.61 ${}^{km^3}/_{s^2}$. Two satellites in similar orbits but different orbital planes can have relative velocities up to 15.22 ${}^{km^3}/_{s^2}$. The order of magnitude of relative velocities in LEO is therefore taken as approximately 15 ${}^{km^3}/_{s^2}$

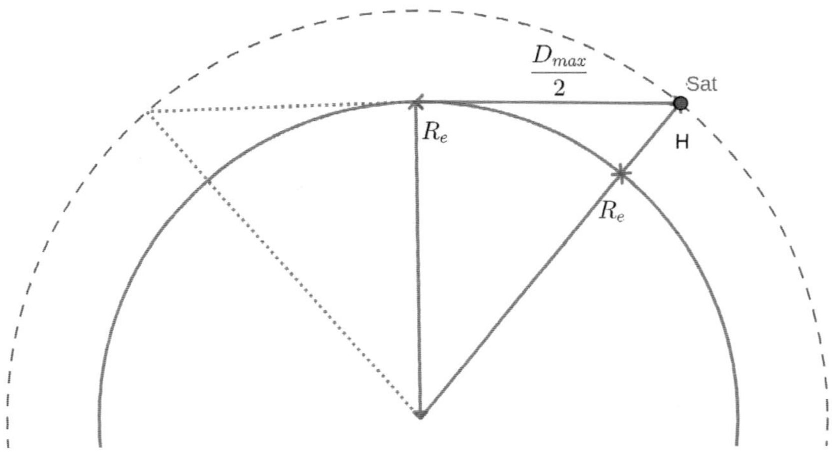

Figure 1: Maximal Line Of Sight Between Two Satellites of the same orbit

For satellites in the same orbital altitude, the maximum Line of Sight (LOS) distance can be geometrically derived (Fig. 1). Half the maximum LOS distance L is given by:

$$\frac{D_{max}}{2} = \sqrt{H(2R_E + H)} \quad (5)$$

where H is the orbital altitude and R_E is Earth's radius. For satellites at 500 km altitude, this yields a total LOS distance of approximately 5148 km. Therefore, the order of magnitude of maximum LOS distances in LEO is taken as approximately 5000 km.

These maximum theoretical values exceed practical detection requirements. The sensor's electrical and physical limitations will constrain the required detection distances.

II.B Optical Requirements Analysis

The optical requirements are derived from the characteristics of the target objects of detection. These space debris particles range in size from $1 \sim 100cm$ characteristic length, yielding a target optical cross-section of anywhere between $1cm^2$ to $1m^2$. Ideally, these targets would be detected and acquired at a distance of approximately $\sim 500km$, with a relative velocity of up to 15 ${}^{km}/_{sec}$.

To assess the capabilities required to address this detection target, we must take a 1 cm^2 target at a 500 km distance. The Fig 2 displays the satellite optical sensor at the middle (left panel) and an enveloping sphere at a 500 km radius. At this distance from the optical sensor, the target subtends a 1 cm^2 area, which translates to a certain surface area at the sensor, corresponding to a pixel. A powerful lens is usually placed between the sensor and the object to enhance and better resolve the image.

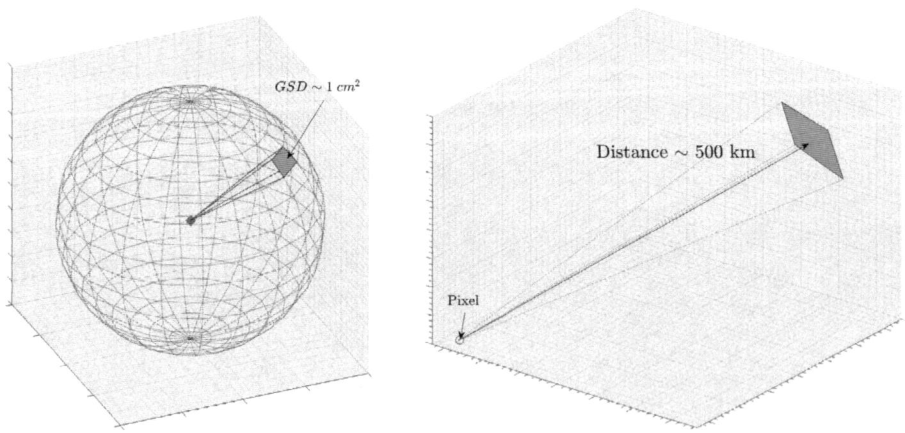

Figure 2: A schematic of pixel/area calculation: on the left, a sphere with the optical device at its center and the surface area it needs to cover. On the right, the surface area of 1 cm^2 is translated to a pixel of a certain size, given the distance and lens characteristics.

The proper way of relating the required pixel-size p to the target size (GSD) A at a distance D is given by:

$$p = \frac{A \cdot f}{D}, \quad (6)$$

in which f is the focal length of the lens.

We require $0 < A \leq 10^{-2}$ m, at a distance of $D \simeq 5 \times 10^5$ m, and our focal length is restricted to the order of several CubeSat units, e,g, $f \lesssim 0.5$ m. With these in mind, we find that the required optical device's pixel characteristic size would be around $p \simeq 10^{-10}$ m. This means a ~ 10 nm-sized pixel. This measure corresponds to a pixel density of around $\sim 4 \times 10^8$ PPI. To drive the point home, the LSST telescope, which has the best pixel density to date among all optical devices, has a pixel size of around 5×10^{-7} m, one order of magnitude short of the task.

If, however, the current trajectory of technological advancement is to be held, this type of resolution might be available as soon as next year if not already employed by state agencies.

Be that as it may, one can relax the target size and/or the required detection distance. For an object of $0.1m$ characteristic length, at $\sim 200km$ distance, we have a pixel size of $\sim 2.5 \times 10^{-7}$, which is technologically possible but currently unfeasible in the context of a space mission. In the following sections, we examine the possible relaxation possible in LEO orbit.

III Methods

As demonstrated in Section II.B, current technology cannot achieve the optimal pixel size required for detecting 1 cm objects at 500 km. This limitation necessitates exploring practical compromises between detection capabilities and technological feasibility. By relaxing either the minimum detectable object size or the maximum detection distance, we can bring the requirements within the realm of current technology while maintaining effective space situational awareness.

The detection distance impacts the sensor's physical implementation, directly affecting the required optical resolution. While larger detection distances warn earlier of potential collisions, they demand larger optics and more sensitive detectors, increasing the sensor's size, mass, and power consumption. These requirements directly conflict with the size and power constraints of small satellites. Therefore, our investigation focuses on finding the minimum viable detection distance that enables reliable collision avoidance.

To evaluate this trade-off between detection distance and estimation quality, measurements were conducted at 500 km, 200 km, and 100 km. At each distance, three parameters were measured:

- Number of successful detections
- Average detection length
- Maximum and minimum detection length

The Keplerian model presented in Section II.A is sufficient for order of magnitude analysis. However, for accurate distance evaluation between satellites, orbital perturbations must be considered. Orbital perturbations are forces that cause deviations from the idealized two-body gravitational motion. In LEO, the primary perturbations include atmospheric drag, which decreases orbital energy over time; Earth's non-spherical gravitational field, particularly the J_2 effect due to Earth's equatorial bulge; third-body gravitational effects from the Sun and Moon; and solar radiation pressure.

These perturbations can be modeled either through numerical integration of high-fidelity equations of motion. This work used the analytical approximations model: Simplified General Perturbations Model (SGP4). SGP4 is a simplified perturbation model analytical model developed for use with the North American Aerospace Defense Command (NORAD) Two Line Elements (TLE) format. It includes secular and periodic perturbations due to Earth's oblateness (J_2 through J_4 terms), solar and lunar gravitational effects, and a simplified atmospheric drag model.

The analysis of detection events required a two-step numerical approach. First, we identified local minima in the relative distance between satellite pairs using the Alfano Negron Close Approach Software (ANCAS) method [13]. This method efficiently locates time periods where the distance between two orbiting objects reaches a local minimum, providing candidate points for potential close approaches. For each identified local minimum, we then employed Newton's method to find the precise times when the relative distance crossed our detection thresholds. Specifically, we solved for the roots of the equation $d(t) - d_{detection} = 0$ where $d(t)$ is the distance between the satellites at time t, and $d_{detection}$ represents our detection thresholds (500 km, 200 km, and 100 km). By applying Newton's forward and backward method, in time from each local minimum, we identified the exact entry and exit times for each detection event. This approach provided precise measurements of detection duration while ensuring we captured all relevant crossing events.

IV Results

We examined six satellite pairs:

- CZ-6A-DEB and STARLINK-1642
- FLOCK-4Y-24 and STARLINK-32398
- IRIDIUM-126 and METEOR-2-5-DEB
- NAPA-1 and STARLINK-31523
- SL-3-RB and STARLINK-2506
- STARLINK-32176 and YAM-3

These satellite pairs were selected from the Satellite Orbital Conjunction Reports Assessing Threatening Encounters in Space (SOCRATES) (Satellite Orbital Conjunction Reports Assessing Threatening Encounters in Space) website [14], which provides conjunction assessments for satellites in LEO.The satellite data was retrieved on 16/12/24. These pairs were chosen for the relatively high collision probability at the time.

The analysis was performed on the selected satellite pairs using a MATLAB® implementation of the SGP4 propagator on an Intel i7-based workstation with 32GB of RAM. For each pair, orbital states were propagated using their respective TLE data, and the relative distance was computed at $0.5s$ intervals over a 14-day period. The propagation results were then filtered using three distance thresholds: 500 km, 200 km, and 100 km. A detection event was defined as a continuous period during which the relative distance remained below the threshold.

For each satellite pair, the results are presented through a pair of graphs and an accompanying table. In the first graph (labeled 'a'), we show the complete distance profile between the two satellites over the 14-day period, providing an overview of their relative motion. The second graph (labeled 'b') focuses on distances below 600 km, with horizontal lines marking our three detection thresholds at 500 km, 200 km, and 100 km. This zoomed view allows for clearer visualization of potential detection events. The accompanying tables quantify these events, presenting three key metrics for each detection threshold: the total number of detection events ("Number of sightings"), the average duration of these events ("Average sighting length"), and their maximum and minimum durations ("Max length" and "Min length" respectively). All duration measurements are provided in days. This presentation format allows us to examine the frequency and quality of potential detection opportunities at different distance thresholds.

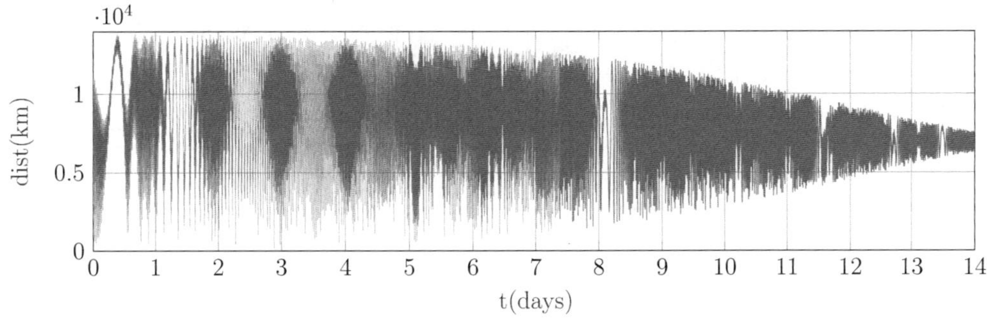

(a) Overview for CZ-6A-DEB and STARLINK-1642

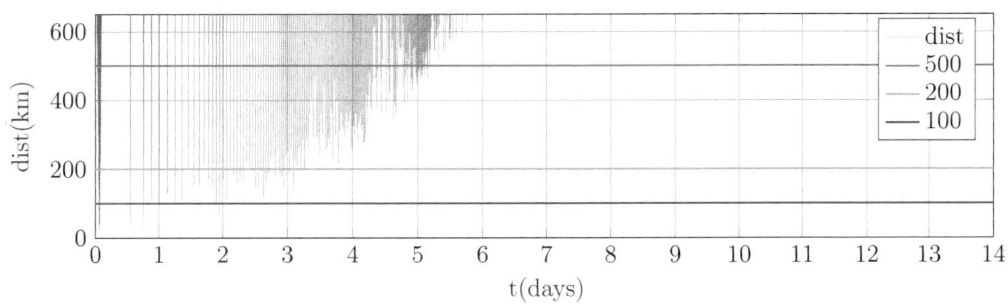

(b) Relative analysis for CZ-6A-DEB and STARLINK-1642

Figure 3: Analysis results for satellites CZ-6A-DEB and STARLINK-1642

Table 1: Tabulated results for satellites CZ-6A-DEB and STARLINK-1642

Detection Distance (km)	Number of sightings	Average sighting lengh (s)	Max length (s)	Min length (s)
500	407	1.1299	2.7051	0.13462
200	63	0.50376	0.91232	0.088804
100	10	0.26388	0.4256	0.16587

The minimum detection lengths recorded in the experiments were approximately 0.1 s, appearing consistently across all measurement distances (100 km, 200 km, and 500 km). With a camera operating at 60 ${}^{frames}/_{s}$, such an encounter could theoretically yield about 10 frames of observation data. Accounting for potential frame drops due to sensor noise,

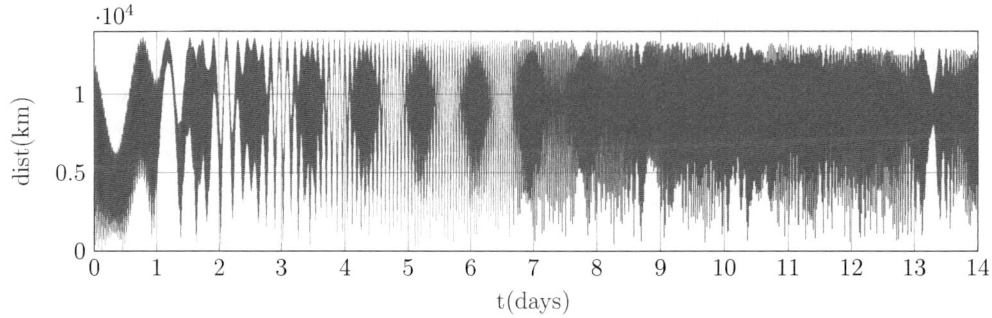

(a) Overview for FLOCK-4Y-24 and STARLINK-32398

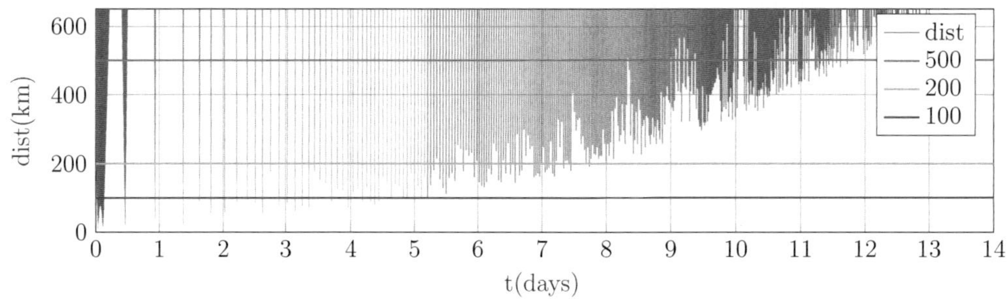

(b) Relative analysis for FLOCK-4Y-24 and STARLINK-32398

Figure 4: Analysis results for satellites FLOCK-4Y-24 and STARLINK-32398

Table 2: Tabulated results for satellites FLOCK-4Y-24 and STARLINK-32398

Detection Distance (km)	Number of sightings	Average sighting lengh (s)	Max length (s)	Min length (s)
500	1190	1.2067	2.3566	0.06633
200	332	0.52893	0.90414	0.035699
100	72	0.26054	0.38021	0.059893

Table 3: Tabulated results for satellites IRIDIUM-126 and METEOR-2-5-DEB

Detection Distance (km)	Number of sightings	Average sighting lengh (s)	Max length (s)	Min length (s)
500	1006	2.1291	20.08	9.3132e-10
200	269	1.0699	9.8747	0.089187
100	64	0.68221	4.7299	0.14262

communication delays, and detection uncertainties in low-signal conditions, a conservative estimate suggests approximately 5 usable frames per encounter. While this represents a limited observation window, it has the potential to be sufficient for initial orbit determination. Each frame provides an angular position measurement of the target object against the star background, and multiple such measurements, even over a short time span, can contribute to trajectory estimation.

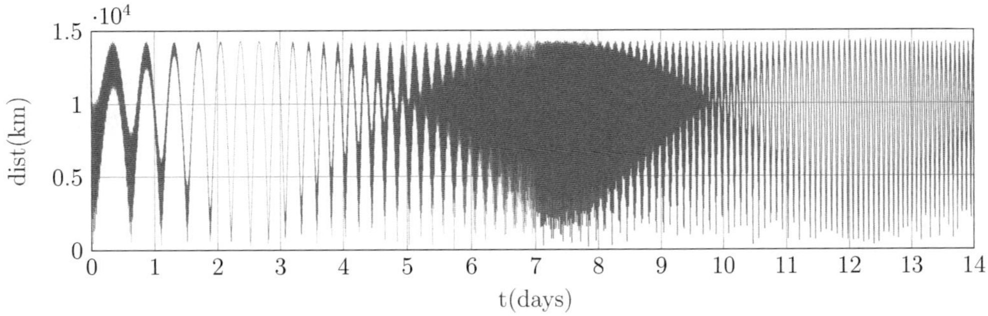

(a) Overview for IRIDIUM-126 and METEOR-2-5-DEB

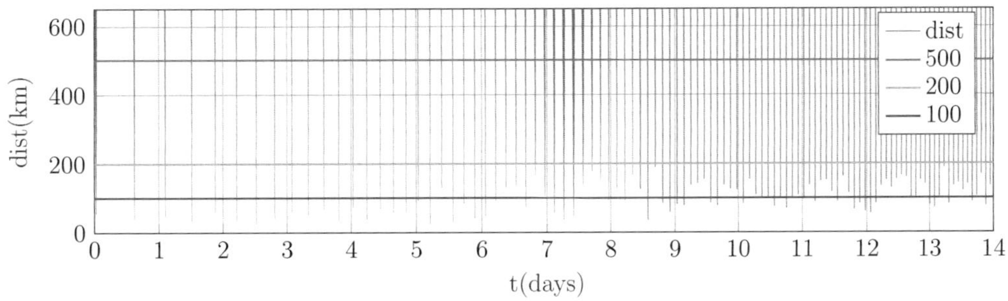

(b) Relative analysis for IRIDIUM-126 and METEOR-2-5-DEB

Figure 5: Analysis results for satellites IRIDIUM-126 and METEOR-2-5-DEB

Table 4: Tabulated results for satellites NAPA-1 and STARLINK-31523

Detection Distance (km)	Number of sightings	Average sighting lengh (s)	Max length (s)	Min length (s)
500	1211	1.268	2.8879	0.11237
200	255	0.56606	1.1431	0.13565
100	36	0.32949	0.55231	0.10147

Table 5: Tabulated results for satellites SL-3-RB and STARLINK-2506

Detection Distance (km)	Number of sightings	Average sighting lengh (s)	Max length (s)	Min length (s)
500	1060	1.2142	2.9074	5.8208e-11
200	254	0.55461	1.158	0.055244
100	51	0.28706	0.57275	0.072893

V Discussion

V.A Sufficient Requirements

Our analysis reveals significant challenges in implementing optimal detection capabilities for space situational awareness on CubeSats. As demonstrated in Section II.B, detecting

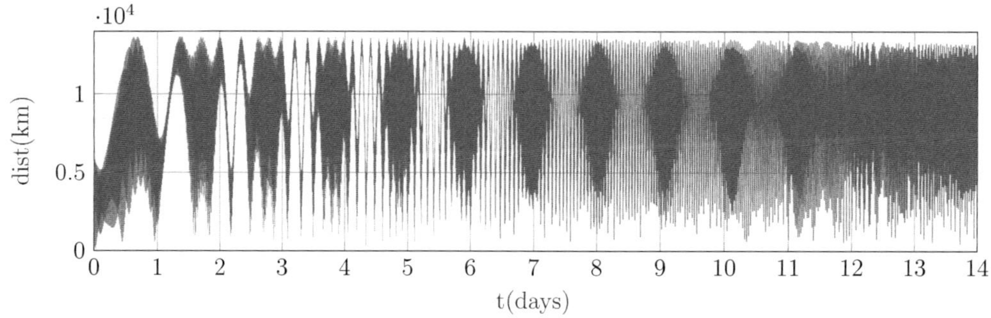

(a) Overview for NAPA-1 and STARLINK-31523

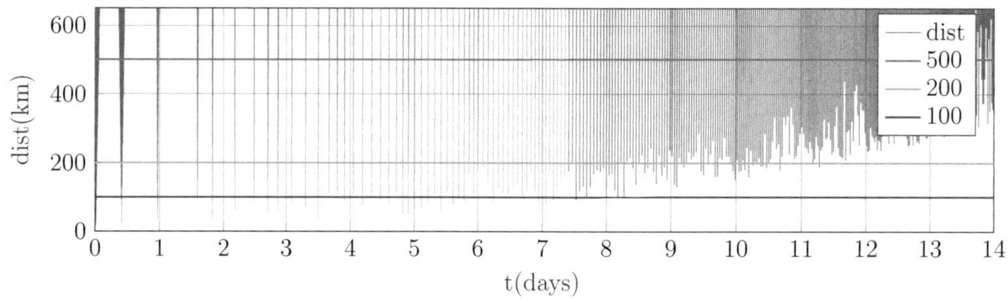

(b) Relative analysis for NAPA-1 and STARLINK-31523

Figure 6: Analysis results for satellites NAPA-1 and STARLINK-31523

Table 6: Tabulated results for satellites STARLINK-32176 and YAM-3

Detection Distance (km)	Number of sightings	Average sighting lengh (s)	Max length (s)	Min length (s)
500	1080	1.1877	2.8487	0.094114
200	236	0.52262	1.1192	0.11298
100	42	0.26447	0.52701	0.15099

1 cm objects at 500 km would require pixel sizes of the order 10 nm, exceeding current technological capabilities by at least an order of magnitude.

The optical requirements become more achievable with current technology by reducing the detection distance to 200 km and focusing on larger objects (approximately 10 cm). However, this compromise significantly impacts detection capabilities. Our results demonstrate that reducing the detection distance from 500 km to 200 km typically decreases the number of detection events by $70 - 80\%$. For example, the FLOCK-4Y-24 and STARLINK-32398 pair reduced from 1190 detection events at 500 km to 332 events at 200 km. Further reducing the detection distance to 100 km makes the optical requirements even more manageable but results in only about $5 - 10\%$ of the original number of detection events.

The duration of detection events also decreases significantly with shorter detection distances. At 500 km, average detection lengths typically range from 1.1 to 2.1 seconds, providing time for detection. However, at 200 km, these durations drop to approximately

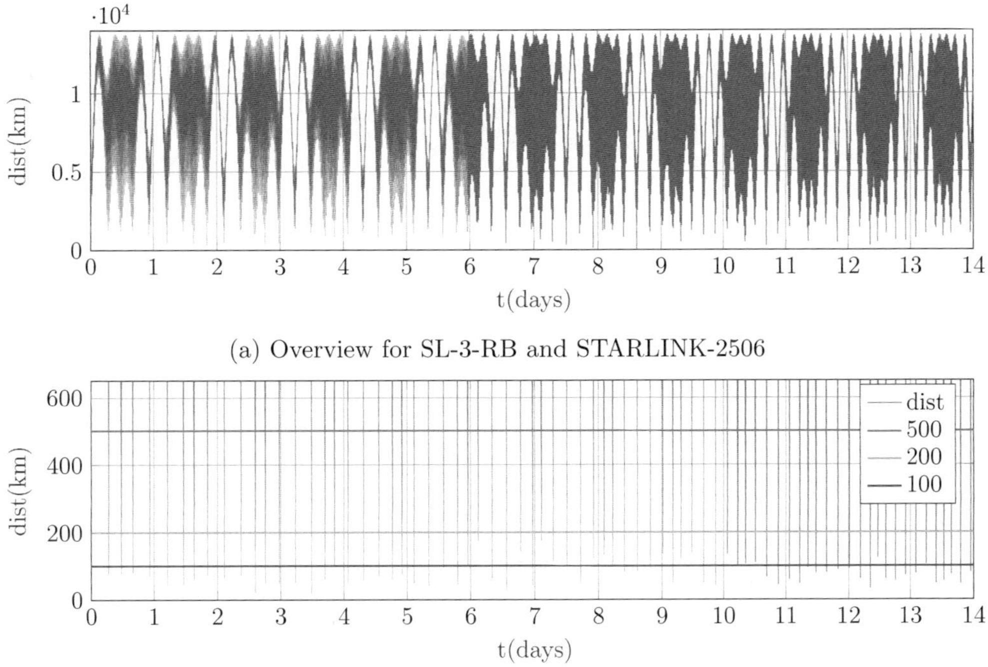

(b) Relative analysis for SL-3-RB and STARLINK-2506

Figure 7: Analysis results for satellites SL-3-RB and STARLINK-2506

$0.5 - 1.0\ s$, and at $100\ km$, they further decrease to around $0.25 - 0.7\ s$. This reduction in observation time constraints the spacecraft's ability to detect threats.

These shorter detection windows have important implications for sensor design. With reduced observation times, particularly at closer detection distances, quickly acquiring and tracking objects becomes crucial.This suggests a need for either wide field-of-view optical systems or sophisticated scanning mechanisms.

V.B Future Work

Future research should examine a broader range of orbital scenarios, including different altitudes, inclinations, and relative velocities between objects. Additionally, studies are needed to determine the minimum detection time required for successful collision avoidance, considering both the computational requirements for orbit determination and the physical constraints of executing avoidance maneuvers. This minimum detection time would help establish more precise requirements for sensor capabilities.

Beyond orbit analysis, effective space situational awareness requires comprehensive monitoring of the surrounding environment. While high-resolution cameras can provide detailed observations, their typically narrow Field of View (FOV) limits their ability to detect potential threats approaching from various directions. A static camera configuration, even with a wide FOV, proves insufficient for the demanding requirements of autonomous collision avoidance in the dynamic orbital environment. Several scanning concepts merit further investigation: gimbaled cameras, which offer maximum flexibility but introduce mechanical complexity; body-mounted scanning, which eliminates additional mechanical systems but impacts spacecraft dynamics; wide-angle optical systems,

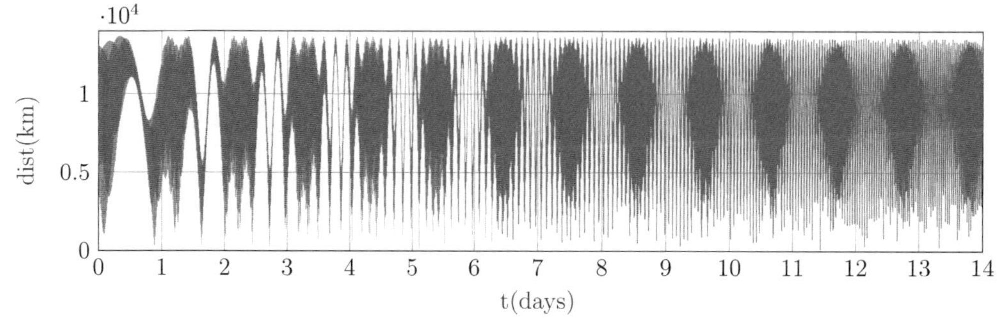

(a) Overview for STARLINK-32176 and YAM-3

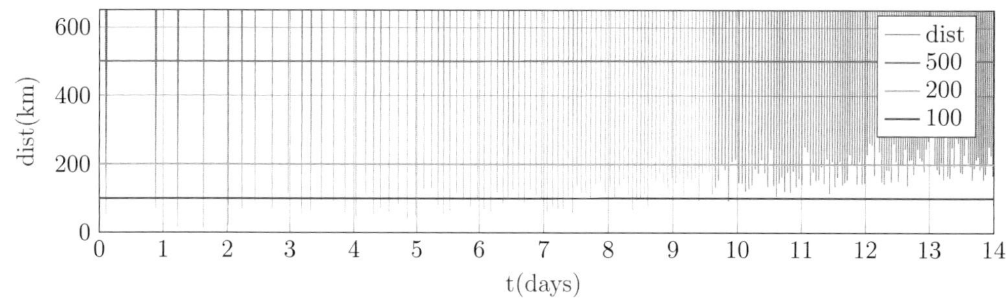

(b) Relative analysis for STARLINK-32176 and YAM-3

Figure 8: Analysis results for satellites STARLINK-32176 and YAM-3

which achieve large spatial coverage but suffer from increased aberrations; and steerable mirrors, which provide a promising compromise between scanning capability and system complexity.

The steerable mirror approach particularly warrants detailed investigation. These systems can achieve scanning capability with minimal mechanical complexity, using a lightweight mirror assembly to redirect the optical path of a fixed camera. Modern fast-steering mirror technologies can achieve high scanning rates with precise control, enabling both systematic surveys and rapid targeting of regions of interest. Future work should focus on optimizing the mirror control mechanisms, establishing specific technical requirements for implementing an effective space situational awareness system within CubeSat constraints, and developing algorithms for efficient scanning patterns that maximize the probability of detection while minimizing power consumption.

References

[1] D. A. Vallado, *Fundamentals of Astrodynamics and Applications*, ch. 4, pp. 247–253. Microcosm Press/Springer, 5 ed., 2022.

[2] F. Letizia, M. Metz, P. Faucher, H. Krag, L. Jing, M. A. Ramos, N. S. Ortiz, T. Schildknecht, and S. Srivastava, "Improving the knowledge of the orbital population new technical means of space debris monitoring," *Acta Astronautica*, vol. 223, pp. 734–740, 2024.

[3] S. Hadji Hossein, M. Acernese, T. Cardona, G. Cialone, F. Curianò, L. Mariani,

V. Marini, P. Marzioli, L. Parisi, F. Piergentili, and F. Santoni, "Sapienza space debris observatory network (sson): A high coverage infrastructure for space debris monitoring," *Journal of Space Safety Engineering*, vol. 7, no. 1, pp. 30–37, 2020.

[4] A. Nocerino, R. Opromolla, G. Fasano, and M. Grassi, "Lidar-based multi-step approach for relative state and inertia parameters determination of an uncooperative target," *Acta Astronautica*, vol. 181, pp. 662–678, 2021.

[5] G. Campiti, M. Tagliente, G. Brunetti, M. N. Armenise, and C. Ciminelli, "Debris detection and tracking through on-board lidar," in *Applications in Electronics Pervading Industry, Environment and Society* (R. Berta and A. De Gloria, eds.), (Cham), pp. 235–241, Springer Nature Switzerland, 2023.

[6] H. M. DOYLE, J. G. TOM, K. H. NOGACEK, C. R. ANDERSON, and J. S. KANG, "Design and development of dragons in-situ orbital debris detection and characterization payload," *TRANSACTIONS OF THE JAPAN SOCIETY FOR AERONAUTICAL AND SPACE SCIENCES*, vol. 59, no. 4, pp. 218–225, 2016.

[7] J. Kuitunen, G. Drolshagen, J. McDonnell, H. Svedhem, M. Leese, H. Mannermaa, M. Kaipiainen, and V. Sipinen, "Debie-first standard in-situ debris monitoring instrument," *EUROPEAN SPACE AGENCY-PUBLICATIONS-ESA SP*, vol. 473, pp. 185–190, 2001.

[8] H. Wilden, C. Kirchner, O. Peters, N. Ben Bekhti, A. Brenner, and T. Eversberg, "Gestra — a phased-array based surveillance and tracking radar for space situational awareness," in *2016 IEEE International Symposium on Phased Array Systems and Technology (PAST)*, pp. 1–5, 2016.

[9] J. Markkanen, M. Lehtinen, and M. Landgraf, "Real-time space debris monitoring with eiscat," *Advances in Space Research*, vol. 35, no. 7, pp. 1197–1209, 2005. Space Debris.

[10] T. Flohrer, H. Krag, H. Klinkrad, and T. Schildknecht, "Feasibility of performing space surveillance tasks with a proposed space-based optical architecture," *Advances in Space Research*, vol. 47, no. 6, pp. 1029–1042, 2011.

[11] D. Spiller, E. Magionami, V. Schiattarella, F. Curti, C. Facchinetti, L. Ansalone, and A. Tuozzi, "On-orbit recognition of resident space objects by using star trackers," *Acta Astronautica*, vol. 177, pp. 478–496, 2020.

[12] J. Kruger and S. D'Amico, "On-orbit performance and lessons learned for autonomous angles-only navigation of a satellite swarm," in *IEEE Space Conference*, Institute of Electrical and Electronics Engineers (IEEE), Jul 2025.

[13] S. Alfano, "Determining satellite close approaches, part ii," *The Journal of the Astronautical Sciences*, vol. 42, no. 2, pp. 143–152, 1994.

[14] T. S. Kelso, "Satellite Orbital Conjunction Reports Assessing Threatening Encounters in Space (SOCRATES)," Dec. 2024. Accessed: 16/12/24 12:05:33UTC.

Satellite Signals-of-Opportunity-Based Navigation: From Simulation to Real-World Implementation

Alex Frid[1], Carlos Caravaca-Gallego[1], Pini Gurfil[1]

[1]Faculty of Aerospace Engineering and Asher Space Research Institute (ASRI), Technion - Israel Institute of Technology, Haifa, Israel

Email: afrid@technion.ac.il, carlosc@campus.technion.ac.il, pgurfil@technion.ac.il

Abstract

Signals of Opportunity (SoOP)-based navigation utilizes satellite constellations, predominantly in low Earth orbit (LEO), to provide positioning, navigation, and timing (PNT) capabilities, offering an attractive alternative or supplement to established Global Navigation Satellite Systems (GNSS). In recent years, LEO constellations, such as Starlink, have emerged as promising SoOP sources. These constellations transmit continuous, stable signals that can be exploited to estimate user position and velocity. By leveraging carrier Doppler measurements and adopting suitable estimation algorithms, accurate navigation solutions become feasible even without direct access to traditional GNSS signals.

This work presents progress in advancing a SoOP navigation framework that transitions from purely simulated environments towards field deployment using actual Starlink transmissions. We begin by outlining a simulation-based methodology that models the Doppler-based navigation solution. The underlying approach employs nonlinear least-squares estimation, supported by the Gauss-Newton algorithm, to jointly solve for receiver position, clock offset, velocity, and clock offset rate. In the simulation, ideal and perturbed conditions are tested to explore achievable accuracies, signal availability sensitivity, and geometric dilution of precision (GDOP) metrics. These simulations incorporate realistic satellite ephemerides, Earth rotation effects, tropospheric delays, and measurement noise models. Under certain conditions, the achieved precision is on the order of meters, and strongly depends on the constellation geometry.

Building on these simulation outcomes, the next step involves implementing a physical receiver architecture to capture and process actual Starlink downlink signals. We describe the receiver's design, including front-end signal conditioning, sampling, and carrier frequency offset tracking. The recorded signals are post-processed using our previously validated algorithms, initially developed and tested in the simulated environment. This hardware implementation aims to confront practical challenges not present in the simulation, such as local oscillator stability, front-end imperfections, interference from other signals, and atmospheric anomalies not perfectly captured in models.

Ultimately, this research seeks to establish the feasibility and reliability of a constellation-based SoOP navigation framework built on Starlink signals.

Manuscript 4 submitted to 64th Israel Annual Conference on Aerospace Sciences (IACAS). Received December 14, 2024.

Electro-Hydrostatic Actuation and Steering System for Nose Landing Gear

Ilan Berlowitz
iberlowitz24@gmail.com

Abstract

The electrification of aircraft plays a crucial role in advancing greener air transportation. To make the "more electric aircraft" a reality, several system-level solutions must be developed to provide airframers with the necessary technologies.

A highly efficient decentralized hydraulic power generation system offers airframers the flexibility to position the hydraulic power pack closer to consumers, eliminating the need for long pipe runs between power generation and usage points. Two sizes of efficient power packs are proposed: one designed for high power demand, typically used in main landing gear retraction and extension, and a mid-sized pack for primary flight control systems in the tail (rudder and elevators), as well as for nose landing gear actuation and steering. These systems operate entirely on electric power, positioning them as key components in the electrification of future aircraft.

Electro-hydrostatic power packs (HEPPs), which are electrically driven and decentralized, supply pressurized fluid on demand. This configuration enables higher pressure levels, weight reduction, and compatibility with environmentally friendly fluids. These self-sufficient systems can be optimized independently from other aircraft systems, enhancing overall efficiency. The baseline pressure is set at 3000 psi, with the capability to scale up to 5000 psi or scale down, providing either continuous or on-demand operation based on future platform requirements.

Electro-motor pumps (EMPs) are the core of the HEPPs. The chosen EMP baseline integrates a permanent magnet synchronous motor (PMSM) and a fixed displacement internal gear pump (IGP).

The advantages of EMPs include high efficiency across the entire flow range, an excellent power-to-weight ratio, adaptability to any electrical power supply, advanced pressure control capabilities, low heat rejection (ideal for small decentralized systems), and reduced hydraulic pulsation, meeting Aerospace Standard AS595E.

Future aircraft architectures are evolving toward greater electrification, requiring systems that can effectively utilize electric power. While pure electro-mechanical actuation has become the standard for secondary flight controls, it is not practical for landing gear due to the significant weight and space constraints imposed by the high-power demands.

To address this, a self-sufficient electro-hydrostatic actuation (EHA) system is proposed for the nose landing gear. This system performs all essential functions, including steering, extension, retraction, unlocking, and door operation, powered entirely by electricity.

The challenges of implementing EHA in landing gear applications include:

1. **Dynamic Loads and Stability**: Nose landing gear (NLG) dynamics involve high-energy events, such as shimmy oscillations[1], which, although infrequent, must not compromise hydraulic power generation or system stability.
2. **Installation Constraints**: The electrical steering, extension, and retraction system comprises multiple components, including the motor-pump, motor control electronics, low-pressure reservoir, and filters. Integrating these elements into a single power pack could result in excessive weight and size, making installation difficult, especially in smaller aircraft. In some cases, it may be necessary to divide the components into smaller units to fit aircraft-specific space constraints.
3. **Maintainability Requirements**: On-wing maintenance for the NLG EHA differs from that of flight control EHAs. For example, replacing retraction actuators will involve opening hydraulic lines. After such maintenance, tasks like on-wing bleeding are necessary to ensure the system is flight-ready, making maintainability a critical design consideration.

Despite these challenges, the proposed solution leverages the high-power density of hydraulic systems while significantly reducing fluid volume. It maintains a conventional hydraulic architecture for essential functions like steering and retraction. This approach minimizes installation impact and mitigates the risks typically associated with new technologies by integrating mature components in an innovative configuration for landing gear applications.

The presentation aims to highlight innovative solutions for advancing aircraft electrification, particularly through the development of decentralized hydraulic power systems and electro-hydrostatic actuation (EHA) technologies. It will showcase how these solutions contribute to greener air transportation by enhancing system efficiency, reducing weight, and increasing compatibility with

---

[1] Shimmy oscillations refer to a type of undesirable, high-frequency vibration that can occur in the nose landing gear (NLG) of an aircraft, typically during taxiing, takeoff, or landing. These oscillations are characterized by rapid, side-to-side movements of the nose wheel. Shimmy can be caused by various factors, such as improper damping, misalignment, or aerodynamic forces, and it can lead to instability or excessive wear on the landing gear components if not properly controlled. Although these oscillations are infrequent, they must be managed to ensure the safety and stability of the aircraft during ground operations.

environmentally friendly fluids. The presentation will also address key challenges in integrating electric-driven hydraulic systems for critical applications, such as landing gear and flight controls, while preserving the power density and reliability of traditional hydraulic systems. Furthermore, it will demonstrate the benefits of modular, self-sufficient power packs and electro-motor pumps (EMPs), emphasizing their role in achieving scalable, efficient, and maintainable systems for future aircraft architectures.

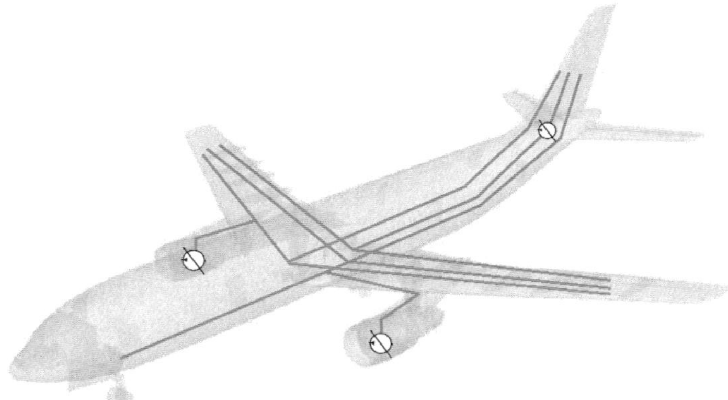

Figure 1: Conventional hydraulic systems

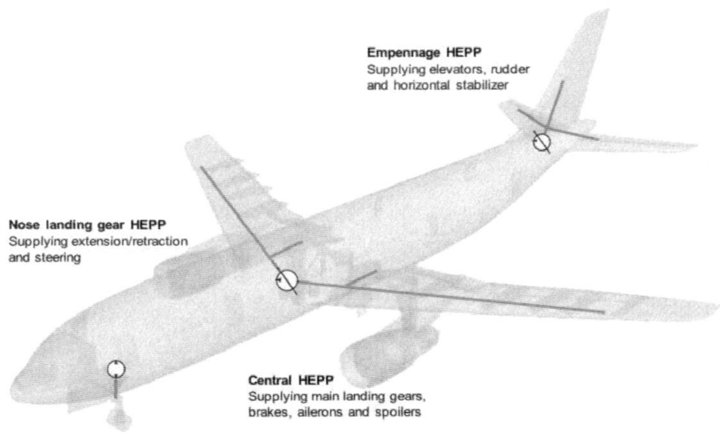

Figure 2: Decentralized hydraulic power generation

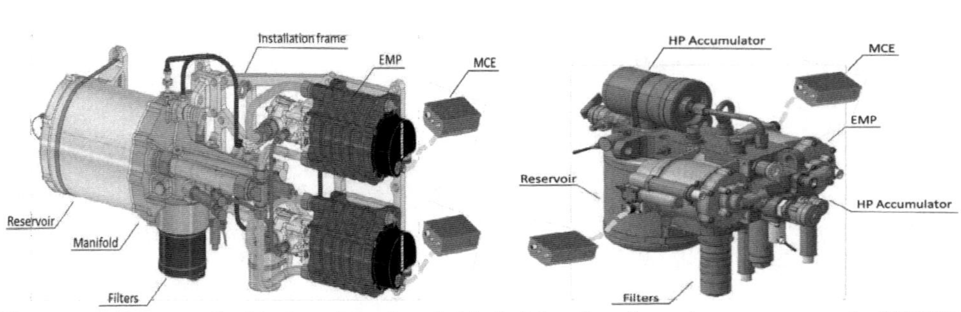

Figure 4: Large (left) & mid-sized (right) hydroelectric power pack (HEPP)

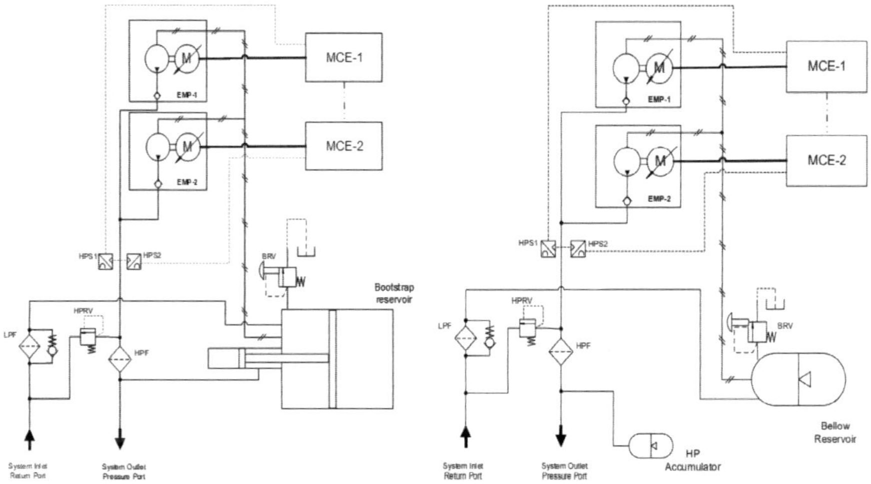

Figure 5: Large (left) & mid-sized (left) HEPP architecture

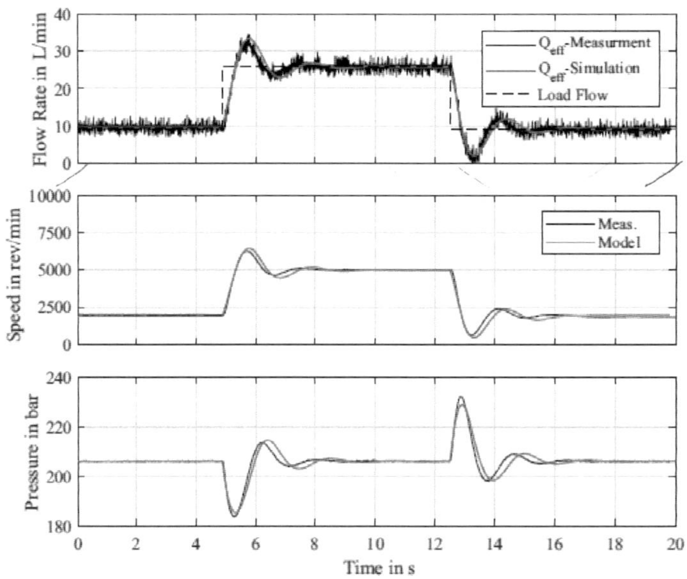

Figure 6: Simulation model of the EMP

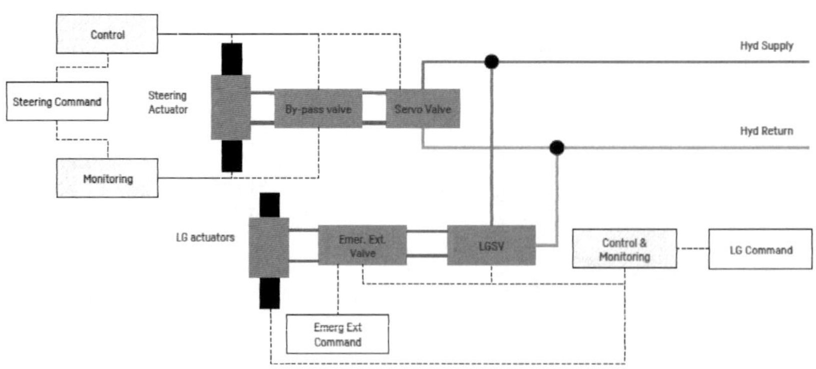

Figure 7: Conventional nose landing gear actuation & steering system

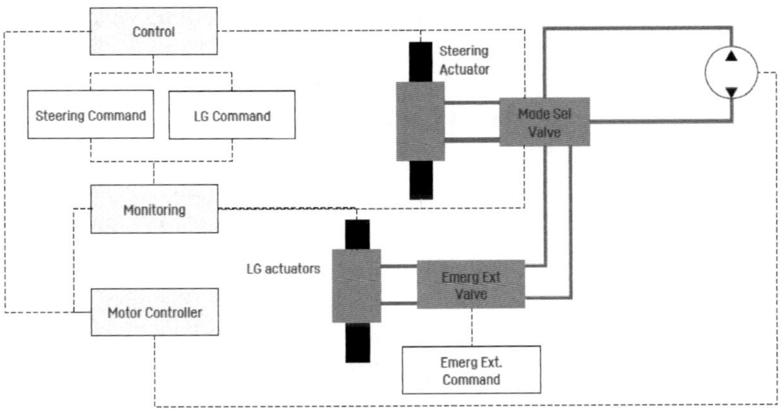

Figure 8: Electrical nose landing gear actuation & steering system

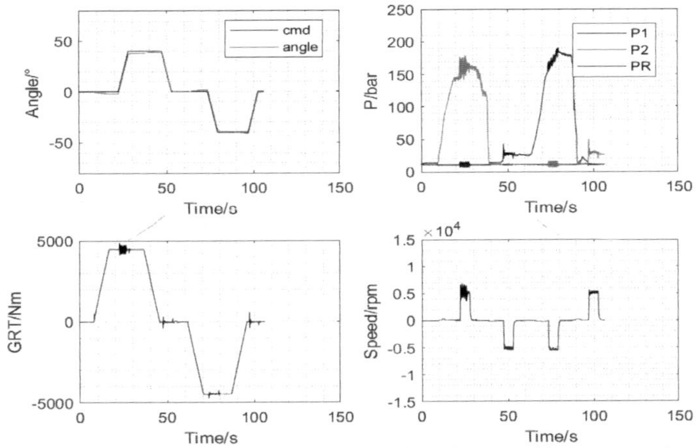

Figure 9: NLG performance under load - medium steering rate

References

- Rosero J A, Ortega E, Aldabas E and Romeral L. *Moving towards a more electric aircraft*. IEEE AE Systems Magazine, 2007.
- Maré J C and Fu J. *Review on signal-by-wire and power-by-wire actuation for more electric aircraft*. Chinese Journal of Aeronautics (CJA), March 2017.
- Aerospace recommended practice 4752B - *Aerospace - design and installation of commercial transport aircraft hydraulic systems*. SAE International, 2013.
- Trochelmann N, Rave T, Thielecke F and Metzler D. *An investigation of electro-hydraulic highly efficient power package configurations for a more electric aircraft system architecture*. Deutscher Luft- und Raumfahrtkongress, Munich, Germany, 2017.
- Van den Bossche D. *The A380 flight control electro-hydrostatic actuators, achievements and lessons learnt*. Proceedings of the 25th International Congress of the International Council of the Aeronautical Sciences (ICAS), Hamburg, Germany, 2006.
- Lima L Z, Duval M N, Zantner T, Metzler D, Lammering T, Thielecke F, Trochelmann N and Bischof P S. *Development of a highly efficient hydraulic power pack*. Proceedings of the 33rd Congress of the International Council of the Aeronautical Sciences (ICAS), Stockholm, Sweden, 2022.
- Guidi G, Amokrane M, Borgarelli N and Malleret F. *Benefits and Challenges of Electro-Mechanical Actuation on Aircraft Landing and Braking Systems* - ATA32 EMAs. More Electric Aircraft Conference, Bordeaux, France, 2021.
- Glas P, Schilling C, Schmelz K. *Aircraft landing gear*. US Patent "9,469,396 B1", July 30th, 2014.
- Duval M N, Bassett, S, Lammering T, *Performance Verification of Electro-Hydrostatic Nose Landing Gear Actuation and Steering*. Proceedings of the 9th International Conference on Recent Advances in Aerospace Actuation Systems and Components (R3ASC), Toulouse, France, 2023.

Comparative Analysis of Aircraft Freighter Conversion Strategies

Ilan Berlowitz
iberlowitz24@gmail.com

Abstract

The global air cargo industry has grown significantly over the past decade, driven by e-commerce expansion, demand for faster shipping, and efficient logistics. As expectations for quick deliveries rise, airlines and cargo operators face pressure to expand fleets and offer cost-effective, flexible solutions. One key strategy to meet these demands is passenger-to-freighter (P2F) conversion, which enhances cargo capacity without the high costs of new freighters.

P2F conversions are a cost-effective and environmentally friendly alternative to purchasing new freighters. By repurposing older aircraft, airlines extend their operational life, reduce costs, and increase fleet flexibility. This approach has become critical as e-commerce fuels rising air cargo demand. Converted freighters support various needs, from long-haul routes requiring high payload capacities to regional routes demanding flexibility and frequent flights. They enable operators to customize fleets for specific requirements, adapting to diverse cargo and route demands.

Freighter conversions provide significant operational benefits beyond cost savings. Airlines can modify aircraft to transport general goods, high-value items, perishable and specialized cargo. Key structural changes include reinforcing fuselages and floors for heavier loads, installing larger cargo doors, and removing passenger amenities. These modifications ensure safety compliance and optimize aircraft for cargo use.

The study highlights differences between long-haul and regional freighter conversions. Long-haul models are designed for substantial payloads and long distances, making them ideal for international routes. In contrast, regional models prioritize fuel efficiency, faster turnaround times, and flexibility for frequent short-distance operations. By examining the strengths and limitations of each, the presentation offers insights into their practical applications across varied operational contexts.

The presentation provides a detailed comparison of popular aircraft freighter conversions, focusing on five key performance metrics: payload capacity, range, fuel efficiency, cargo handling systems, and turnaround time. These metrics are analysed to choose the most suitable freighter model for cargo transport needs.

Long-Haul Freighters

- *Mammoth Freighters 777-200LRMF/-300ERMF*: - Engineered to align closely with Boeing's production 777F, these conversions feature a rear extra-large 163.2" x 120" main deck cargo door (XL-MDCD), enabling efficient loading of oversized cargo while maximizing flexibility and payload capacity. The 777-200LRMF offers extended range. The 777-300ERMF, with its longer fuselage, provides higher cargo volume, optimizing capacity for high-density freight. Both models deliver high payload capability and extensive range, making them ideal for intercontinental routes connecting major global hubs.

- *Boeing 767-300BCF (Boeing Converted Freighter)*: - Balancing payload capacity and fuel efficiency, this versatile model suits medium to long-haul routes. Supported by Boeing's direct engineering expertise, ensuring reliability and seamless integration into existing fleets.
- *Airbus A330-200/-300P2F*: - The Airbus fly-by-wire (FBW) freighter family offers advanced technology and high performance. The A330-200P2F is optimized for higher-density freight and longer-range performance, while the longer-fuselage A330-300P2F is particularly well-suited for integrators and express carriers, thanks to its high volumetric payload capability for lower-density cargo. Both A330 conversions provide exceptional payload capacity, long-range efficiency, and versatile cargo space with up to 26 main deck positions. Known for their fuel efficiency, these models are a sustainable option for intercontinental operations, making them ideal for express and general freight markets.

Regional Freighters

- *Boeing 737-800BCF and AEI 737-800SF*: - Equipped with 5 supernumerary seats, these freighters enhance crew capacity and support special cargo operations. Designed for fuel efficiency, they are ideal for short-haul routes with frequent schedules, particularly in dense domestic markets.
- *Precision Aircraft Solution A321PCF and Elbe Flugzeugwerke EFW and ST Engineering A320/A321P2F*: - Featuring a containerized lower deck, these freighters maximize payload capacity with 14 main-deck pallet positions and 10 lower-deck containers, making them ideal for e-commerce and express operations. Offering +28% more volume and +22% higher payload compared to the 737-800F. They provide 4 + 2 crew configurations. These models are designed for dense cargo handling and extended range, ensuring greater operational flexibility. The A321P2F can carry up to 28 tons of payload, and with its longer fuselage compared to the A320P2F, it provides 28% more volume, catering particularly to the growing express and e-commerce markets. The A321P2F stands as the most viable replacement for the Boeing 757 freighter.

Figure 1: Long-haul models Mammoth Freighters 777-200LRMF/-300ERMF (left), Boeing 767-300BCF (center) and Airbus A330-200/-300P2F (right)

Figure 2: Regional models Boeing 737-800BCF (left), AEI 737-800SF (center-left), Precision Aircraft Solutions A321PCF (center-right), EFW and ST Engineering A320/A321P2F (right)

Boeing and Airbus have distinct approaches to production and freighter conversions, reflecting their differing strategies, priorities, and product designs. Here are some key differences:

Focus on Freighter-Specific Models

- **Boeing:**
  - Boeing has historically focused heavily on the cargo market, offering purpose-built freighters like the 747-8F, 767-300F, and 777F.
  - Boeing also supports Passenger-to-Freighter (P2F) conversions for older aircraft, such as the 737, 747, and 767, through programs like the Boeing Converted Freighter (BCF) initiative.
  - The designs prioritize maximizing cargo volume, payload, and operational efficiency, leveraging decades of experience in the cargo industry.
- **Airbus:**
  - Airbus has less extensive experience in freighter-specific models but has developed cargo variants like the A330-200F and the upcoming A350F[1].
  - Airbus' P2F conversions are relatively newer, focusing on A320, A321, and A330 models, often through partnerships with specialized companies (EFW joint venture with ST Engineering).

Structural Design Considerations

- **Boeing:**
  - Boeing aircraft are often designed with the freighter market in mind, even for passenger models. For instance:
    - The 747 was initially designed to be easily converted into a freighter.
    - The 777 is optimized for a cargo door and a spacious main deck.
  - Boeing's designs often accommodate significant cargo payloads, with reinforced floors and structures to handle heavier loads.

---

[1] The Airbus A350F features an advanced cargo door designed for optimal loading efficiency. With a cut-out dimension of 4445 mm (175 inches) and a clear opening of 4305 mm (169.5 inches), it surpasses the nose and side doors of the Boeing 747F. This door is 15% wider than the one of the Boeing 777F, allowing for the seamless one-go entry of large items. Specifically built to transport all new large engines, the A350F accommodates oversized cargo without the need for disassembly. This makes it an ideal choice for the aviation industry, setting a new standard in the freighter market with its superior door dimensions and enhanced operational efficiency.

- **Airbus:**
  - Airbus designs are primarily passenger-focused, with freighter capabilities added later.
  - For conversions, Airbus must modify structures more extensively, such as reinforcing floors and adding large cargo doors, especially for models like the A320/A321P2F.

Market Segmentation and Strategy

- **Boeing:**
  - Boeing targets a broader range of freighter sizes, from narrowbody (737-800BCF) to large widebody freighters (777F, 747-8F).
  - Boeing also leverages its strong ties with cargo operators like FedEx, UPS, and DHL.
- **Airbus:**
  - Airbus focuses more on medium-range and narrowbody conversions like the A320/A321P2F, which is suited for e-commerce and regional freight markets.
  - Airbus emphasizes fuel efficiency and the integration of modern technologies, like the A350F, which meets stringent environmental regulations.

Conversion Partnerships

- **Boeing:**
  - Boeing handles many freighter conversions through its in-house programs and trusted partnerships, ensuring a seamless process.
- **Airbus:**
  - Airbus relies heavily on partnerships, such as the EFW joint venture, to manage its P2F conversion programs. These collaborations enable Airbus to expand its freighter offerings while leveraging external expertise.

Technology and Efficiency

- **Boeing:**
  - Boeing's freighter conversions and dedicated freighters focus on durability and reliability, often favoring proven technologies to minimize costs for cargo operators.
- **Airbus:**
  - Airbus emphasizes sustainability and fuel efficiency, particularly for new models like the A350F, which comply with ICAO's CO_2 standards.

Summary

The future of Boeing and Airbus freighters, along with freighter conversions, will be shaped by advancements in sustainability, XL-MDCD, payload capacity, automation & digitalization and human factors. Key to this evolution will be enhancing the integration between aircraft structures and the additional margins used in various fittings, particularly where cargo structure and equipment intersect with the original OEM framework. With the continued rise of e-commerce and global trade, demand for versatile, fuel-efficient, and flexible freighter solutions - across both narrowbody and widebody aircraft - will increase significantly.

Table 1: Widebody Freighter Conversions Technical Specifications

Model	Mammoth 777-200LRMF/-300ERMF	Boeing 767-300BCF	Airbus A330-200/-300P2F
Payload Capacity	102 - 106 tons	57 tons	61 - 63 tons
Range	Up to 9,200 km	6,300 km	Up to 7,700 km
Fuel Efficiency	Optimized for large payloads with improved fuel efficiency	Efficient for medium-haul operations	Superior in its class
Cargo Handling	Optimized for oversized cargo, including special cargo configurations	Designed for standard pallet and containerized cargo	Flexible configurations for mixed cargo
Turnaround Time	Moderate to long, due to larger size and cargo capacity	Quick, adaptable for various loads	Moderate, balanced with efficiency

Table 2: Narrowbody Freighter Conversions Technical Specifications

Specification	Boeing 737-800BCF	AEI 737-800SF	Precision A321PCF	EFW A320/A321P2F
Payload Capacity	23.9 tons	23.6 tons	27 tons	27 - 28 tons
Range	Max 3,685 km	Max 3,700 km	Max 3,400 km	Max 3,500 km
Fuel Efficiency	Fuel efficient for medium-haul routes with improvements in fuel burn	Slightly optimized for shorter routes	Optimized for medium-haul, improving fuel consumption	High efficiency for medium-haul routes
Cargo Handling	Palletized deck	Flexible loading configurations	Additional pallet positions and optimized cargo handling	Flexible upper and lower deck configurations for diverse cargo types
Turnaround Time	~1 hour (optimal conditions)			

References

- Long-Haul Recommendations: *The 777-200LRMF for maximum payload capacity or A330-300P2F for slightly better fuel efficiency.*
- Regional Recommendations: *The A321P2F and A321PCF for higher payload and fuel efficiency.*
- Boeing. (2023). *Boeing 777-200LR Freighter Specifications.* Retrieved from https://www.boeing.com
- Airbus. (2023). *Airbus A330-300P2F Freighter Conversion Data Sheet.* Airbus Commercial Aircraft Technical Publications.
- Smith, J., & Lee, R. (2022). *Trends in aircraft freighter conversions: Meeting the global cargo demand.* Aviation Week & Space Technology, 180(5), 48-53.
- Martin, L. (2023). *Fuel efficiency in freighter conversions: A comparative study of narrow-body models.* Journal of Air Transport Management, 62, 123-134.
- Oliver Wyman. (2023). *Global Fleet & MRO Market Economic Analysis 2023.*
- IBA Group. (2023). *Freighter Conversion Market Insights and Forecast.* Ascend by Cirium.
- Doe, A., & Campbell, S. (2022). *Operational efficiencies of converted freighter aircraft in regional markets.* In Proceedings of the AIAA Aviation Forum (pp. 112-125).
- Young, P., & Carter, E. (2021). *Evaluating turnaround time and cargo handling in freighter conversions.* In Proceedings of the International Cargo Logistics Conference (pp. 56-63). Cargo Logistics Association.
- United Nations Environment Program. (2021). *Environmental impact of freight transport: A global review.* UNEP Transport and Environment Reports.
- Brown, T. (2022). *Case study: Adoption of the 777-200LRMF by global cargo operators.* Cargo Facts, 44(3), 28-35.
- Wilson, K. (2023). *Airline adoption of A321 freighter conversions: A performance review.* Air Cargo News, 37(2), 15-20.

DEVELOPMENT OF REGIONAL AIRCRAFT COMPOSITE LEADING EDGE FOR BIRD STRIKE SUSTAINABILITY - A METHODOLOGY REVIEW

Y. Angel, A. Sawday, E. Ejgenberg
IAI, Ben Gurion International Airport 70100 Israel

INTRODUCTION

The increasing demand for environmentally sustainable aviation has spurred the development of innovative, lightweight, and high-performance materials and structures. Our research, conducted within the EU-funded "Clean Aviation" framework, under HERWINGT project [1] and aims to contribute to the European Green Deal and support climate neutrality by 2050. Specifically, our focus is on addressing the critical issue of bird strikes through advanced simulation techniques designed to optimize aircraft leading-edge designs. The Clean Aviation initiative seeks to develop disruptive technologies that will achieve net greenhouse gas (GHG) reductions of at least 30% compared to the 2020 state-of-the-art, significantly contributing to aviation's sustainability goals

As a whole the HERWINGT project aims for a Hybrid-electric propelled aircraft, based on hydrogen or batteries solutions for regional flights and ultra-efficient aircraft design, utilizing thermal engines suitable for the adoption of sustainable aviation fuels (SAF). This approach is expected to trigger a technology revolution targeting climate-neutral aviation in Europe by 2050. Within the project a novel induction based de-icing system is developed and integrated, thus the methodical process for bird strike substantiation involving multidisciplinary challenges of integrating this novel system is provided in this abstract.

Figure 1: Hybrid Electric Regional Wing Integration Novel Green Technologies

METHODOLOGY and ANALYSIS

Traditional approaches to mitigating bird strike risks often involve heavy, reinforced materials that compromise both fuel efficiency and environmental performance.
To overcome this challenge, we have developed a simulation-based methodology that enables the accurate prediction of bird strike impacts on lightweight leading-edge structures. This framework integrates advanced computational techniques to model the dynamic interaction between birds and aircraft components. By incorporating detailed bird parameters, material properties [2] , and impact mechanics, our model enhances the precision of predicting how lightweight materials behave during bird strikes.

Manuscript 32 submitted to 64th Israel Annual Conference on Aerospace Sciences (IACAS). Received December 31, 2024.

Our simulation model was rigorously validated against experimental data [3] and compared with conventional design approaches. The results demonstrate the potential of this innovative approach to not only improve aircraft safety and resilience but also support the broader objectives of the Clean Aviation initiative. By reducing the weight of leading-edge components, we anticipate improvements in fuel efficiency, reduced emissions, and a lower overall environmental impact, aligning with the aviation industry's push for greener, safer solutions.

This research, part of a collaborative effort led by Airbus Defense and Space, alongside Hellenic Aviation Industries, the University of Patras, and Israel Aerospace Industries, represents a significant step toward leveraging simulation technologies to enhance both aircraft safety and sustainability. Our findings offer valuable insights for future aircraft designs and regulatory frameworks, helping to shape the next generation of aviation technologies.

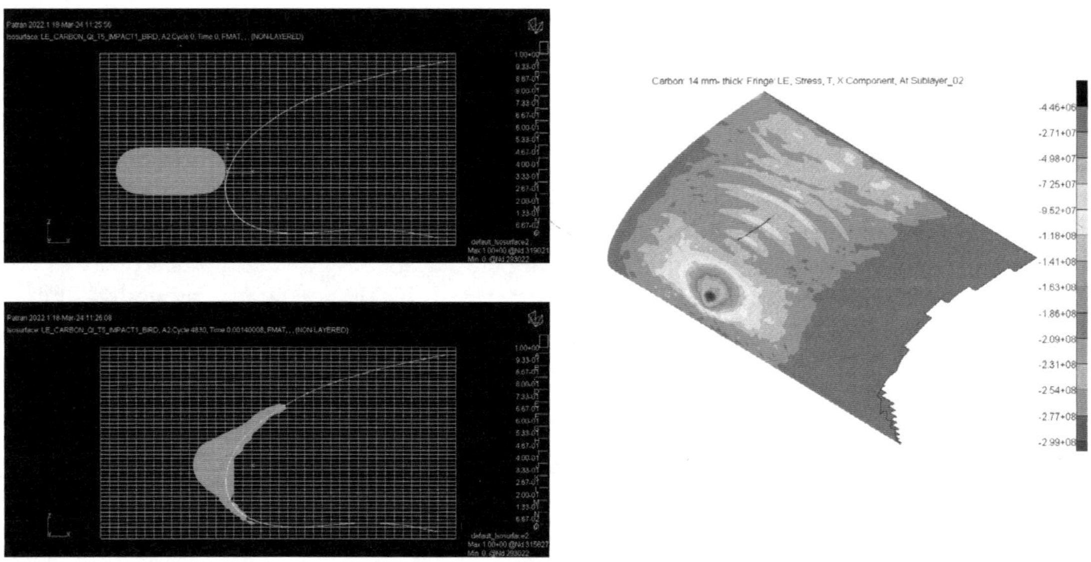

Figure 2: Leading Edge FSI Analysis

REFERENCES

[1] https://herwingt-project.eu/sites/default/files/2023-04/HERWINGT_Poster_outlined.pdf

[2] Ledford. N, Imbert. M, May. M, " High-rate in-plane shear testing of CFRP using the Split Hopkinson Tension Bar", 2021, AIAA

[3] Ramos Valle. L, "Numerical Simulation of Impact Bird Against Rigid Flat Plate and Rigid Edge for Bird Splitting Analysis", 2015, Bachelor's Thesis

Manuscript 32 submitted to 64th Israel Annual Conference on Aerospace Sciences (IACAS). Received December 31, 2024.

Helicopter Engine's Power Estimation Using Regular Flight Data

Liron Darhi[1,2], Ariel Dvorjetski[2], Yehudit Aperstein[1]

[1]Afeka Academic College of Engineering, [2]Israel Air Force

Abstract

Accurately estimating helicopter engine power is essential for ensuring flight safety, optimizing mission performance, and improving maintenance practices. Traditional methods, such as Maximum Power Checks (MPCs), rely on dedicated maintenance flights to calculate the Engine Torque Factor (ETF), a key performance indicator. While effective, these procedures are resource-intensive and infrequent, limiting their practical applicability for continuous monitoring. This study introduces a novel machine learning-based framework for estimating ETF values using routine flight data, significantly reducing the reliance on MPCs.

The proposed methodology leverages Health and Usage Monitoring System (HUMS) data from AH-64A helicopters of the Israeli Air Force. Based on the assumption that engine power remains stable between maintenance events and faults, a nonstandard labeling strategy is employed to annotate regular flights with ETF values derived from preceding MPCs. To ensure data quality, the framework applies advanced preprocessing tools, including a steady-state filter to isolate stable operating points and a consolidator to eliminate redundant data points.

The machine learning models, including XGBoost and CatBoost, were trained using features such as altitude, airspeed, engine torque, gas generator speed (Ng), power turbine speed (Np), and turbine gas temperature (TGT). The models achieved R^2 values of up to 0.94 and RMSE values as low as 0.006, demonstrating robust predictive capabilities across diverse operational ranges. Cross-validation results demonstrated the model's stability and generalization, with training and test RMSE curves showing consistent alignment across steps of a boosting algorithm. Error analysis revealed a normal distribution of the errors, with no identifiable patterns, and minor errors aligning with the RMSE results.

These results validate the feasibility of continuous, data-driven engine power monitoring, providing actionable insights for predictive maintenance, mission planning, and trend analysis. The methodology highlights how modern machine learning techniques can enhance maintenance procedures and power estimation, offering several key advantages: it enables enhanced maintenance planning by reducing the need for frequent power check flights, improves operational efficiency with near real-time (post-flight) insights into engine performance, offers a deeper understanding of engine degradation through power trend analysis over time, and supports informed decision-making regarding operability, maintenance, and replacement scheduling.

Manuscript 72 submitted to 64th Israel Annual Conference on Aerospace Sciences (IACAS). Received January 9, 2025.

Collision Coalescence Study of Impinging Spray Jets.

Ariel Sharon[1] and Yeshayahou Levy[1]
[1]Faculty of Aerospace Engineering, Technion – Israel Institute of Technology, Haifa – 3200003, Israel

ABSTRACT

The present study investigates droplet dynamics using vibrating mesh piezoelectric atomizers (VMA) in an opposing pair jets configuration, to simulate collision-coalescence behaviors observed in cloud environments. The aim is to enhance the understanding of droplet interactions and growth mechanisms in clouds, focusing on the collision-coalescence process critical to raindrop formation. While gravitational collision-coalescence drives growth for droplets larger than 80 µm, the rapid growth of droplets in the 30–80 µm range remains poorly understood, a challenge referred to as the "condensation-coalescence bottleneck."

The research examines droplets ranging from 5 to 50 µm in size, using VMA to explore their growth dynamics. VMAs are particularly suitable for aerosol studies due to their ability to produce repetitive and narrow size range droplets. To replicate cloud-like collision-coalescence, a head-to-head impinging spray configuration was implemented, creating a stagnation zone, to study droplet size growth.

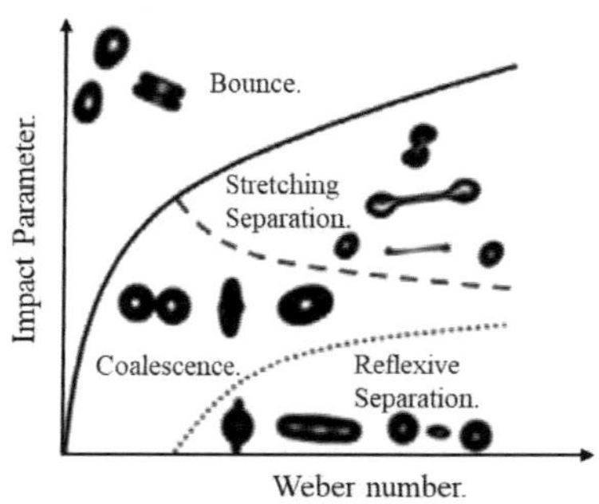

Figure 1 Schematic of the geometry of a collision for water droplet. *NOTE: Absence of bouncing at Impact factor of 0 is unique to water droplets at low diameter.*

For water droplets, where surface tension is significant, collisions result in one of four outcomes:

1. **Bounce/Rebound**: Droplets collide but fail to merge due to gas pressure between them.
2. **Coalescence**: Droplets merge at higher velocities and favorable impact angles.
3. **Stretching Separation**: Any further increase of velocity leads to high-impact energy causing the droplets to elongate and break.
4. **Reflexive Separation**: High energy droplets deform into flattened disks but retract and separate due to surface tension.

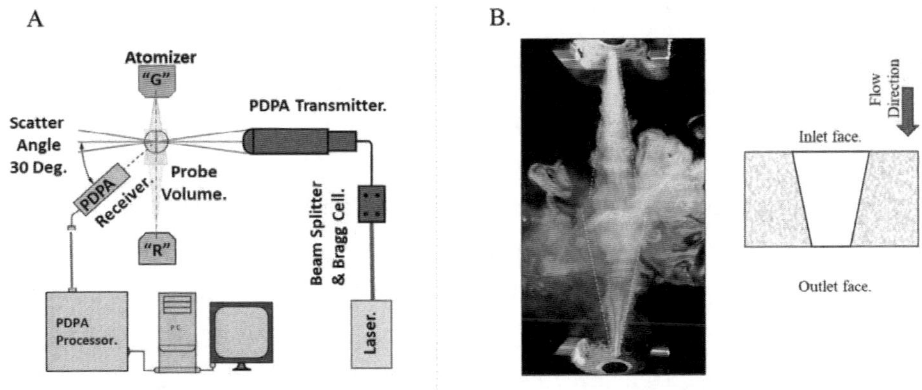

Figure 2 (Right) System Layout – 'VMA G' and 'VMA B' pair. (Left) Light sheet imaging of spray core and microscopic representation of VMA sectional view of one of its multiple nozzles.

This study identifies parameters influencing droplet size growth through collision-coalescence, particularly the collision Weber number and collision Ohnesorge number. Experiments show that droplets smaller than 15–20 μm rarely coalesce between themselves, regardless of velocity. Droplet coalescence was most frequently observed between larger droplets having low velocity (low weber number) introduced with a cloud of high droplet concentration (irrespective of cloud droplet size). In Figure 3, the D_{32} trend for two standalone nozzles (B with a max D_{32} of 22.2 μm and G with 13.4 μm) and the impinging configuration layout of the B+G nozzles, where the max D_{32} of 25.1 μm is displayed. The existence of larger droplets in the stagnation region presents clear evidence of the local occurrence of collision coalescence.

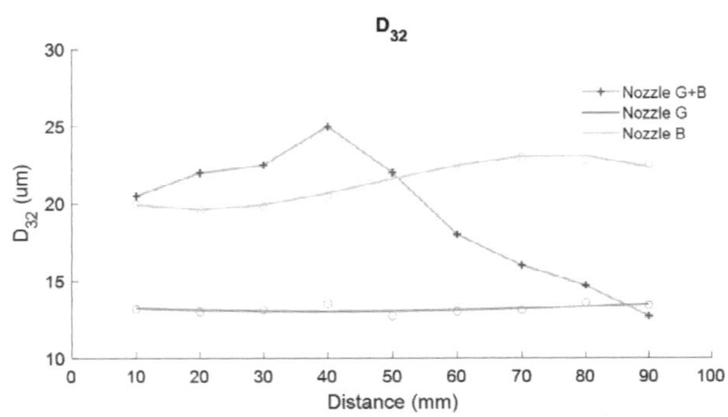

Figure 3 Trend of D32 along central line for Nozzle G, Nozzle B as standalone and in Impinging layout.

Using the measured results, we were able to calculate the Weber number of the droplets along the jets. Interestingly, higher We values were recorded downstream, towards the stagnation plane, indicating that the increase in droplet diameter overcomes the reduction in velocity of the coalesced droplets. This suggests that a faster droplet is colliding with slower upstream droplets traveling in the same direction.

These findings provide insights into the transition from diffusion-driven to collision-driven droplet growth, improving our understanding of precipitation processes.

Optimizing the Performance of Savonius Vertical Axis Wind Turbines for Use in Urban High-Rise Buildings and Rural Locations

By

Moshe Zilberman

Azrieli College of Engineering, Jerusalem, 91035

moshezi@jce.ac.il

Wind energy has experienced rapid growth, as one of the most common and preferred renewable energy sources. Over the past decade, the global annual growth rate of both installed wind power capacity and generated wind energy has been approximately 30%. Two main types of wind turbines are used to harness wind energy and convert it into electricity: Horizontal Axis Wind Turbines (HAWTs) and Vertical Axis Wind Turbines (VAWTs).

HAWTs are primarily installed in large-scale wind farms in rural areas or offshore locations, where the wind flow is strong and consistent. However, in urban and suburban environments, the wind flow is typically more turbulent, intermittent, and variable due to changes in surface roughness and upstream obstructions such as buildings, trees, and other large structures. VAWTs are more desirable and better suited for urban settings, where the wind flow is less predictable. By capturing wind from all directions, VAWTs do not require complex yaw mechanisms, as they are not highly sensitive to wind direction. VAWTs are attractive due to their lower noise levels and reduced maintenance costs; with fewer moving parts, installation becomes easier and not expensive. They are the better option for off-grid installations in both on-ground and high-rise rooftop buildings.

The work, presented here, is a development of a new Savonius wind turbine, featuring two blades with a bi-polynomial shape. This innovative turbine was down-selected from thirty different Savonius VAWT models tested using the 30x30 wind tunnel in Azrieli. It was then compared with five other models, including the original Savonius design, and demonstrated significant improvement over all the compared models [3].

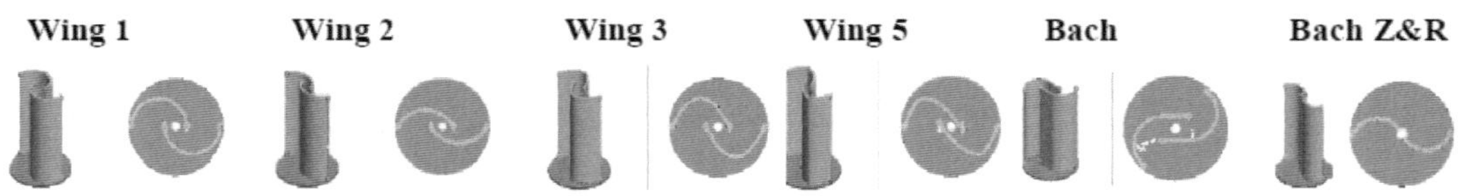

Figure 1. The Selected Wing 5 and the other five types of Savonius Wind Turbines

Three-step procedures were employed in testing the various types of VAWTs:

1. Measuring the rotational speed ω as a function of wind velocity V_∞ 2. Static measurements of three aerodynamic coefficients, lift (C_L), drag (C_D), and torque (C_T) 3. Dynamic measurements of torque and power coefficients, (C_T) and (C_P), respectively, versus Tip Speed Ratio (λ)

$$C_T = \frac{T}{0.5\,\rho V_\infty^2 SR} \quad ; \quad C_p = \frac{P}{0.5\,\rho V_\infty^3 S} = C_T\lambda \tag{1}$$

Manuscript 40 submitted to 64th Israel Annual Conference on Aerospace Sciences (IACAS). Received January 5, 2025.

Wing 5 was derived from a Myring-defined aerodynamic shape. Yet, instead of a symmetrical quadratic formula, it was modified to a bi-polynomial shape, with its maximum value shifted from 50% to 70% radius distance from the turbine axis. Compared with the other models, it exhibited the best torque and power coefficient (see Fig. 2).

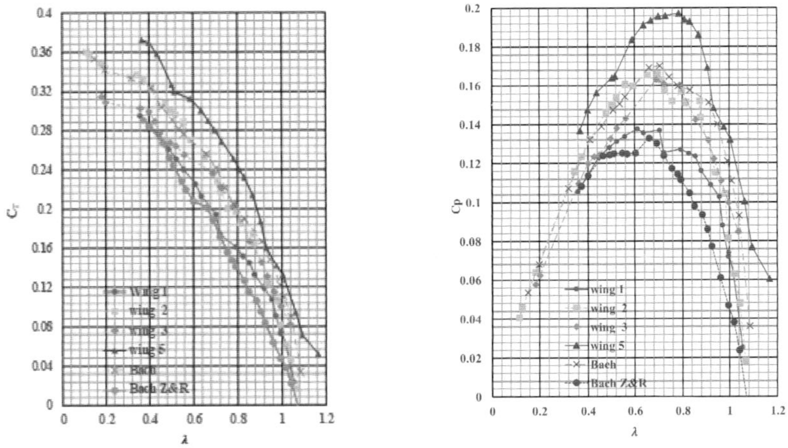

(a) Torque coefficient (b) Power coefficient

Figure 2. C_T and C_p as a function of λ for different VAWTs

It is shown that due to the relatively low Reynolds number in our tests, by similarity, the absolute power coefficient value is expected to be between 0.25 to 0.3.

The atmospheric boundary layer wind speed distribution was used to demonstrate profitable electricity production when the developed wind turbine shape is used on rooftops of high-rise buildings.

Simplified modeling of constrictor plasma at the Technion arc heated wind tunnel

The arc-heated wind-tunnel at the Technion is used to test materials for thermal protection systems for hypersonic vehicles. The heating electrical arc is generated inside of a segmented constrictor which is connected to a converging-diverging nozzle, and then to a test chamber. It is important to understand the environment inside the test chamber to better analyze the material testing results. For this purpose, it is also important to characterize the plasma generated in the constrictor. In this talk we present the radial temperature distribution of the constricted plasma calculated using a simplified model based on a heat diffusion equation. The inputs to the calculation are the measured current through, and voltage across, the constrictor. The degree of equilibrium between heavy and light plasma species was also calculated. It was found that radiation losses play an important role in the calculation and cannot be neglected as in other smaller laboratory arcs with lower temperatures. It was also found that light and heavy plasma particles are in a high degree of equilibrium.

Hydrogen Generation from Water and Aluminum with different additives

Elinor Kostjukovsky and Alon Gany

Faculty of Aerospace Engineering, Technion - Israel Institute of Technology

Hydrogen is considered the most environmentally friendly alternative fuel due to its high energy release and its eco-friendly oxidation byproducts—water or water vapor. However, due to storage and transportation issues, hydrogen is not employed much today. Hydrogen can be produced on site and on demand through the chemical reaction between aluminum and water, offering a high storage capacity and does not require high-pressure hydrogen storage:

$$Al + 3H_2O \rightarrow Al(OH)_3 + \frac{3}{2}H_2$$

The primary problem with this reaction is that it does not happen under normal circumstances, because the aluminum particles naturally create an oxide layer that prevents further reaction. A patented thermal-chemical process, developed at the Fine Rocket Propulsion Center, Faculty of Aerospace Engineering of the Technion has overcome this issue by activating aluminum with 2.5% of LiH [1]; the process involves dissolving the lithium hydride activator in an anhydrous solvent and depositing it on the aluminum particles, thereby disrupting the protective properties of the oxide layer. Parametric studies on the reaction between water and activated aluminum for hydrogen production were carried out by Rosenband and Gany [2] and Elitzur et al [3].

The objective of this study is to investigate the option of using a mix of aluminum and the LiH activator without applying the activation procedure and to compare the results to those of activated aluminum (Fig. 1). In addition, we will study the use of alternative additives which are more readily available (e.g., $Ca(OH)_2$ and NaOH) and are less expensive than LiH. Through these experiments, we want to identify the optimal conditions (e.g., water/aluminum ratio, reaction temperature) for maximizing hydrogen yield and improving

Manuscript 65 submitted to 64th Israel Annual Conference on Aerospace Sciences (IACAS). Received January 9, 2025.

reaction efficiency. Concentration of the hydroxyl (OH) at the onset of the reaction and the pH during the reaction have been monitored as well. The results shown in Fig. 2 demonstrate that the reaction of aluminum mixed with $Ca(OH)_2$ and NaOH yields similar hydrogen output to that of aluminum mixed with LiH, suggesting that these alternative activators may offer comparable efficiency for on-demand hydrogen production.

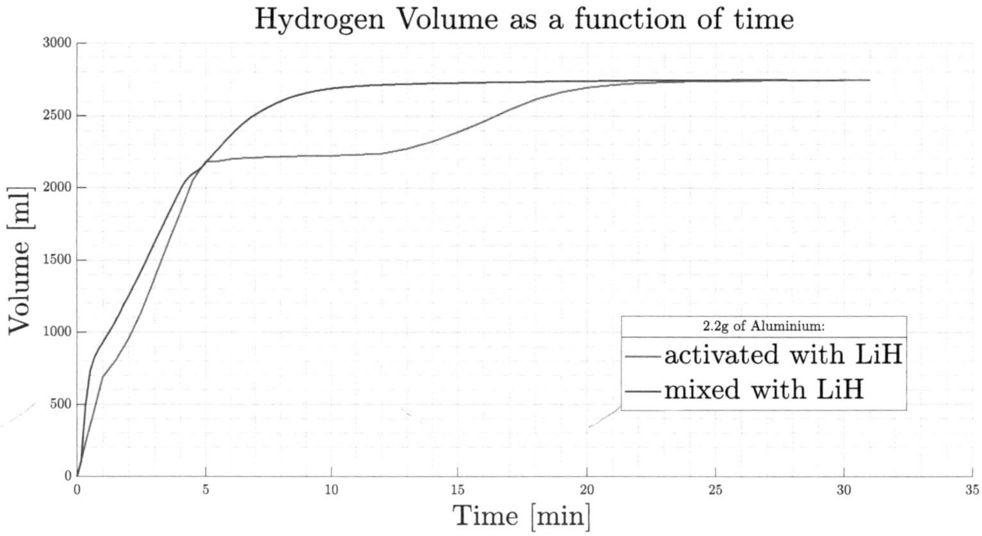

Figure 1: Aluminum powder activated with LiH vs. Aluminium powder mixed with LiH at the same ratio and conditions: 10 ml of tap water at the same temperature, Al mass 2.2 gram.

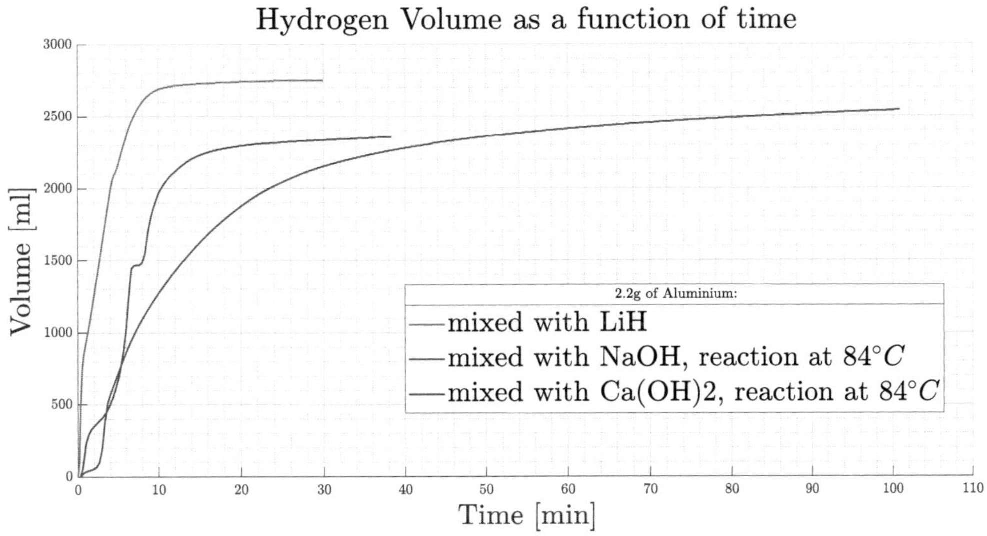

Figure 2: Aluminium with different additives with the same amount of water, the reaction with LiH was at initial room temperature whether the reactions with Ca(OH)2 and NaOH, Al mass 2.2 gram.

References

1. Rosenband, V. and Gany, A., "Compositions and Methods for Hydrogen Generation", US Patent 8,668,897 B2, March 11, 2014.
2. Rosenband, V. and Gany, A., "Application of Activated Aluminum Powders for Generation of Hydrogen from Water", *International Journal of Hydrogen Energy*, Vol. 35, Issue 20, Oct. 2010, pp. 10898-10904, ISSN 0360-3199.
3. Elitzur, S., Rosenband, V., and Gany, A., "Study of Hydrogen Production and Storage Based on Aluminum-Water Reaction", *International Journal of Hydrogen Energy*, Vol. 39, 2014, pp. 6328-6334.

Lattice Boltzmann method simulations of the laminar flow in a two-dimensional double cavity configuration

Itamar Blumenfeld*, Hadar Ben-Gida†
Technion Israel Institute of Technology, Haifa 3200003, Israel

Two opposing open cavities are referred to as a double cavity configuration. The flow field through this configuration is a well-documented phenomenon observed in various contexts, including sudden expansions and contractions in pipe flows, as well as the exit of confined jets. In this configuration, the dynamic interaction between the two shear layers emerging from each wall of the inlet channel is significantly influenced by both the cavity and the channel dimensions. This study investigates the characteristics of a two-dimensional laminar and incompressible flow within the double cavity configuration, utilizing an in-house Lattice Boltzmann Method (LBM) code. The LBM is an evolving and well-documented methodology rooted in kinetic theory, with numerous versions available for implementation. The in-house LBM code employs either a Bhatnagar-Gross-Krook (BGK) or a two-relaxation-time (TRT) model as the collision operator, the bounce-back (BB) method as an obstacle boundary condition, the Zou-He scheme for domain boundary conditions, and the D2Q9 discrete velocity set. The code is first validated with two prominent and extensively documented cases. The first case involves the flow over a two-dimensional circular cylinder, while the second concerns the flow around a two-dimensional square cylinder positioned within a channel. In both validation scenarios, we examined the pressure coefficient distribution surrounding the respective obstacles (circular or square), as well as the resultant lift and drag coefficients and the Strouhal number associated with vortex shedding occurring behind the obstacle. A good agreement was observed between the numerical results obtained with the LBM code and the reference results utilized for comparison. The double cavity configuration studied herein has a cavity length-to-hight ratio (L/H) of 2 and a channel width (D/L) of 0.2 and was studied for a range of cavity length-based Reynolds numbers (Re_L) from 3600 to 10000. The results were compared to existing literature and analyzed to assess the impact of the Reynolds number on the laminar flow regimes in the double cavity configuration.

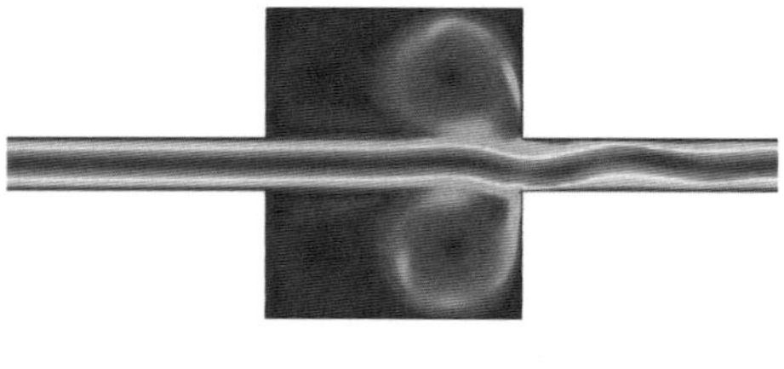

(a)

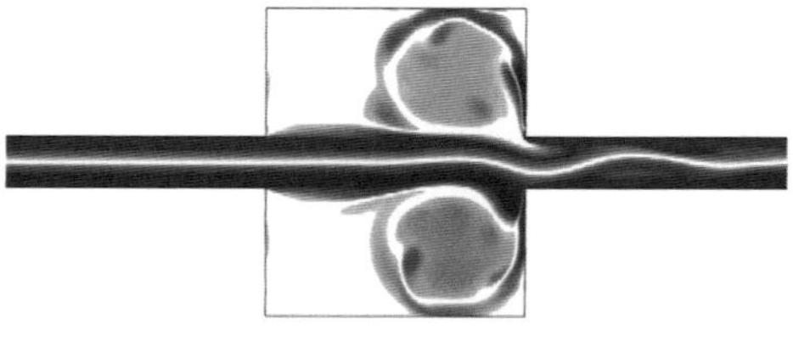

(b)

Fig. 1 Instantaneous flow field of the double cavity configuration for $Re_L = 7000$, $L/H = 2$ and $D/L = 0.2$. (a) Velocity magnitude. (b) Vorticity.

*Undergraduate Student, Faculty of Aerospace Engineering,itamar.b@campus.technion.ac.il.
†Research Fellow, Faculty of Aerospace Engineering, bengida@technion.ac.il.

Wake measurements of heavy vehicle rear-end models with custom-designed passive flow control devices

Niv-Haim Mizrahi
Supervisors: Prof. Oksana Stalnov and Prof. Alexander Liberzon
Meadow Aerodynamic Laboratory, Tel-Aviv University

Abstract
Our study investigates the role of passive flow control devices with corner configurations on the aerodynamic performance of a heavy vehicle's rear end. The research focuses on wake flow features to explain drag reduction and enhance efficiency. Experiments were performed in a low-turbulence wind tunnel at the Meadow Aerodynamic Laboratory, Tel-Aviv University. Using a 1:9 scale truck wind tunnel model equipped with serval rear-end configurations: baseline (no attachment), 2D upper wing, and advanced 3D corner configurations surrounding the rear end of the truck with the aerodynamic profile. A 7-hole probe system, coupled with a pressure scanner and linear motors, was calibrated to capture three-dimensional flow characteristics. Drag force measurements, using a load cell, were conducted alongside velocity field mapping to provide comprehensive insights into the flow dynamics.

Initial findings indicate that 3D attachments significantly alter the wake structure, reducing aerodynamic drag compared to the baseline and 2D models. The drag coefficient was reduced across all tested configurations, with the "Clean Connection", a smooth connection between the sides and the top sides, and "Vertical Flap", a rear profile similar to "H" shape, demonstrating the most pronounced effects. These reductions result from decreased flow interference and enhanced reattachment behind the truck. The measured Reynolds is defined as, $Re = \frac{uL}{\nu}$, where L defined by the root of the surface area of the truck, u, the velocity in the flow direction and ν is the kinematic viscosity. The drag coefficient is defined, $C_d = \frac{2F_d}{\rho u^2 A}$, where F_d is the measured drag force, ρ is the mass density and A is the truck surface area. All examined configurations reduce the drag coefficient compared to the baseline, with 3D configurations most effective. The "Clean" achieved the greatest drag reduction 12% - 18% at Reynolds number of $3.2 - 6.5 \times 10^5$, with the decrease was especially significant at higher Reynolds numbers while the drag decrease percentage of the 2D upper profiles remained steady at 9% compared to the baseline measured drag, the baseline measured drag coefficient ranges between 0.615 – 0.594 from lower to higher Reynolds. The experimental study verifies the critical role of rear-end geometry in shaping the flow characteristics, decreasing the drag coefficient while changing the wake flow according to the rear-end geometry. The wake measurements indicate that the decrease in the drag results from the flow adhesion to the different rear-end attachments minimizing the turbulent flow phenomena and reducing the aerodynamic drag.

Manuscript 77 submitted to 64th Israel Annual Conference on Aerospace Sciences (IACAS). Received January 12, 2025.

Abstract for the 64-th IACAS 2025 Meeting

Linear Stability of Complex Compressible Cavity Flows

Vojtěch Pezlar*

Czech Technical University, Prague, Czech Republic

Marlon S. Mathias†

Universidade de São Paulo, São Paulo, Brasil

Vassilis Theofilis‡

Technion - Israel Institute of Technology, Haifa, Israel

and Marcello Medeiros§

Universidade de São Paulo, São Paulo, Brasil

A comparative study of global linear stability using the multi-domain matrix-forming approach and matrix-free time-stepping approach is done on open cavities within the compressible continuum framework. Global stability of steady subsonic[1] and supersonic[2] 2D baseflow is investigated by employing the BiGlobal linear stability analysis. The continuum framework baseflow for both approaches is obtained using a second-order accurate, finite-volume based DNS solver. Both approaches allow for the investigation of stability on both cartesian (rectangular cavities (Fig. 1a)) and curvilinear[3] (cavities with modified downstream walls (Fig. 1b)) domain geometries.

The multi-domain matrix-forming method utilizes a high-order compact finite difference discretization scheme[4] together with an iterative, sparse, parallel Arnoldi solver to perform a LNSE-based stability analysis. The time-stepping approach does not require the Jacobian matrix to be explicitly computed; instead, it uses the flow solver itself as part of its algorithm.

*Doctoral Student, Faculty of Fluid Mechanics and Thermodynamics, CTU

†Professor, Department of Mechanical Engineering, USP

‡Professor, Departement of Aerospace Engineering, Technion

§Professor, Department of Aerospace Engineering, USP

Manuscript 55 submitted to 64th Israel Annual Conference on Aerospace Sciences (IACAS). Received January 8, 2025.

It is based on the Arnoldi method of solving eigenproblems, which also does not require the matrix itself, just the ability of multiplying it by an arbitrary vector. The method is rearranged in such a way that simulating the temporal evolution of a base flow plus some small disturbance for a short period of time is equivalent to multiplying the Jacobian matrix by this disturbance. Some details on the time-stepping implementation have been presented,[5] while both methods will be discussed during the conference.

(a) Vertical wall

(b) Modified downstream wall

Figure 1: Comparison of open-cavity downstream wall configurations

References

[1]Bres, G. A. and T., C., "Three-dimensional instabilities in compressible flow over open cavities," *Journal of Fluid Mechanics*, Vol. 599, mar 2008, pp. 309–339.

[2]Sun, Y., Taira, K., Cattafesta, L. N., and Ukeiley, L. S., "Biglobal instabilities of compressible open-cavity flows," *Journal of Fluid Mechanics*, Vol. 826, sep 2017, pp. 270–301.

[3]Das, S. and Cohen, J., "Reducing Pressure Oscillations by Change in Rear-Face Geometries at Subsonic and Transonic Speeds," *AIAA Journal*, Vol. 55, No. 12, dec 2017, pp. 4401–4411.

[4]Paredes, P., Hermanns, M., Le Clainche, S., and Theofilis, V., "Order 104 speedup in global linear instability analysis using matrix formation," *Computer Methods in Applied Mechanics and Engineering*, Vol. 253, jan 2013, pp. 287–304.

[5]Mathias, M. S. and Medeiros, M. A. F., "Optimal Computational Parameters for Maximum Accuracy and Minimum Cost of Arnoldi-based Time-Stepping Methods for Flow Global Stability Analysis," *Theoretical and Computational Fluid Dynamics*, Vol. 36, No. 6, Dec. 2022, pp. 1013–1036.

Manuscript 55 submitted to 64th Israel Annual Conference on Aerospace Sciences (IACAS). Received January 8, 2025.

Instability and transition of the flow over a surface gap

Victor B. Victorino[1], Felipe O. Aguirre[2] and Marcello A. F. Medeiros[3]

Abstract— **This paper presents experiments carried out in a low-turbulence wind tunnel to investigate the effect that gaps produce on the boundary layer stability. The model has a variable-depth cavity system inserted into a flat plate. The results show four different regimes. The first, which occurs for shallow gaps, is marked by an interaction with the boundary layer viscous instability (T-S wave). In the second regime, a turbulent mixing layer inside the cavity rapidly decays downstream but leaves a residual amplitude in the T-S wave. In the third regime, a strong Rossiter mode causes an abrupt transition at the gap. At even deeper gaps the oscillations in the gap reduce substantially.**

I. INTRODUCTION

Surface gaps disturb the boundary layer and push the laminar-turbulent transition upstream or immediately trip a turbulent boundary layer. [1] and [2] report the so-called tripping limit or bypass transition bound, with dependence on the gap geometry relative to the displacement thickness. [1] and [2] also model the impact on the transition through ΔN, a jump in the transition amplification factor N. Furthermore, the flow over the gap produces instability related to the mixing layer [3] and the highly separated flow inside the cavity [4]. The Rossiter mode is caused by feedback between acoustic waves and the vortices in the mixing layer. Also, the centrifugal mode, associated with the recirculation in the gap, becomes unstable. These modes are governed by cavity dimensions, boundary layer thickness, Reynolds, and Mach number.

II. EXPERIMENTAL SETUP

Experiments were conducted in the University of Sao Paulo's Low Acoustic Noise and Turbulence wind tunnel (LANT). The LANT has a test section of 1 m $\times$ 1 m $\times$ 3 m, a maximum speed of 42 m/s and turbulence intensity of 0.03% (empty). Measurements used an A.A. Lab System AN-1002 anemometer sensed by a Dantec 55P15 probe. The transition was detected through a Preston tube. A 24-bit NI acquisition system recorded the fluctuation signal. The flat plate model with a chord of 2320 mm has a variable-depth cavity located 600 mm from the model's leading edge, capable of reaching in-depth. All the edges are sharp. The cavity spans 600 mm and has a 30 mm width in the streamwise direction.

[1]Victor B. Victorino is with the Sao Carlos School of Engineering, Department of Aeronautical Engineering, University of Sao Paulo, Av Trabalhador sao-carlense - 400, Sao Carlos/SP, Brazil barcelos.victorino@usp.br

[2]Felipe O. Aguirre is with the Sao Carlos School of Engineering, Department of Aeronautical Engineering, University of Sao Paulo, Av Trabalhador sao-carlense - 400, Sao Carlos/SP, Brazil felipe.aguirre@sc.usp.br

[2]Marcello A. F. Medeiros is with the Sao Carlos School of Engineering, Department of Aeronautical Engineering, University of Sao Paulo, Av Trabalhador sao-carlense - 400, Sao Carlos/SP, Brazil marcello@sc.usp.br

III. RESULTS

Figure 1 exhibits contours of RMS of the velocity fluctuation (log-scale) measured in the gap near-field. There is an increase in RMS as the gap becomes deeper, including a sharp increase that is close to the bypass transition boundary established by [2] (dashed lines). The streamwise evolution of the disturbance velocity just below the bypass transition exhibited a turbulent spectrum in the mixing layer but underwent a fast relaminarization downstream. Moreover, the transition location did not change significantly with the gap depth confirming the deep-gap limit region indicated by [2]. The bypass transition was caused by a Rossiter mode identified in the spectrum as a sharp peak for deeper gaps. For even deeper gaps the oscillations in the gap decayed.

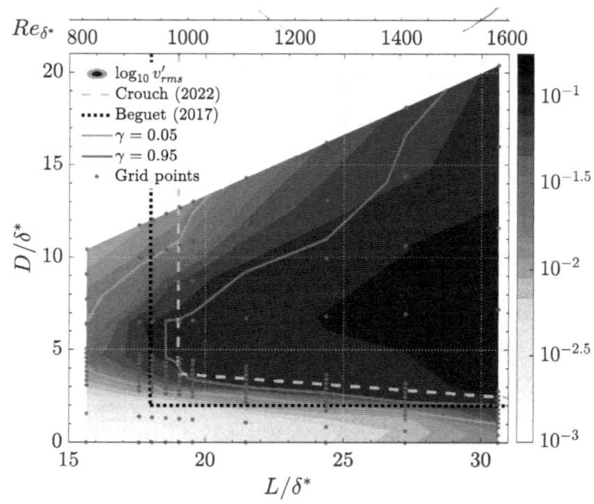

Fig. 1. RMS contours of disturbance velocity measured in the near field for several gap geometries.

REFERENCES

[1] S. Beguet et al., Modeling of transverse gaps effects on boundary-layer transition, Journal of Aircraft, 54 (2), 2017.

[2] J. Crouch et al., Characterizing surface-gap effects on boundary-layer transition dominated by Tollmien–Schlichting instability, Flow, 2 (E8), 2022.

[3] J. E. Rossiter, Wind-tunnel experiments on the flow over rectangular cavities at subsonic and transonic speeds, Reports and Memoranda No. 3438, 1964.

[4] G. A. Brés and T. Colonius, Three-dimensional instabilities in compressible flow over open cavities, Journal of Fluid Mechanics, 599 pp. 309-339, 2008.

Manuscript 28 submitted to 64th Israel Annual Conference on Aerospace Sciences (IACAS). Received December 24, 2024.

Mach effect on instability and transition of the flow over a surface gap

Felipe O. Aguirre[†], Paulo C. V. Paino[◇] , Hadar Ben-Gida[‡] and Marcello A. F. Medeiros[⋆]

Abstract— **This study investigates the effect of surface gap instability on boundary layer transition using Direct Numerical Simulation (DNS). Global stability analysis and DNS reveal reduced oscillation amplitudes as one moves from transonic to supersonic regimes (Mach 0.9–3.0). Rossiter, centrifugal, and Mack modes instabilities are analyzed. Results highlight the influence of the Mach number on flow dynamics, providing insights into the open cavity behavior and boundary-layer bypass transition mechanisms at high-speed flows.**

I. Introduction

Surface gaps are two-dimensional cavities that interact with boundary layer viscous instability (T-S mode) or can immediately trip a turbulent boundary layer. Beguet et al. [1] and Crouch et al. [2], who studied gap-induced transition, found that the tripping limit or bypass transition bound depends on the gap geometry relative to the displacement thickness at the gap's leading edge. They also model the transition caused by the gap and its impact.

Additionally, the flow over the gap generates instabilities related to the mixing layer [3] and the highly separated flow inside the cavity [4]. Rossiter modes arise from the feedback mechanism between the acoustic waves and vortices in the shear layer. Furthermore, centrifugal modes in the gap can also become unstable. In the supersonic regime, the flow presents Mack modes, unstable acoustic modes crucial in early supersonic boundary layer transition. All these modes are influenced by Mach, which is the focus of the current study work.

II. Methodology

In this study, we utilize Direct Numerical Simulation (DNS) combined with linear stability analysis (LST) of the time-stepping kind. Both codes were developed in-house and are open-source[1]. The DNS solves the compressible Navier-Stokes equations (NSE) using a Cartesian, structured, and stretched mesh. Buffer zones dissipate oscillations near open boundaries by modifying node spacing and the spatial derivative order of accuracy. Spatial terms are approximated using a spectral-like differentiation scheme, and a spatial anti-aliasing filter is applied. Temporal integration employs a fourth-order Runge-Kutta method. The flat plate is maintained at an isothermal condition. For the LST analysis, selective frequency damping is optionally applied to obtain a steady base flow for globally unstable cases. Parallel execution is handled via pencil and slab domain decomposition.

III. Results

The Mach number (Ma) tested were 0.9, 1.2, 2.0, and 3.0, all with a displacement thickness-based Reynolds number of $Re_{\delta^*} = 734$. The non-dimensional gap's depth and length were $D/\delta^* = 6.11$ and $L/\delta^* = 12.22$, respectively. Figure 1 shows the time variation of the streamwise velocity fluctuations downstream of the gap for different freestream Mach numbers. Results indicate a reduction in oscillation amplitudes as Ma moves from the transonic to the supersonic regime. The flow seems to be highly unstable at $Ma = 0.9$ and 1.2. At $Ma = 2.0$, the instability reduces, and at $Ma = 3.0$, it is marginally unstable. Linear stability analysis (LST) is being conducted to confirm these observations and investigate the reason for this stabilization.

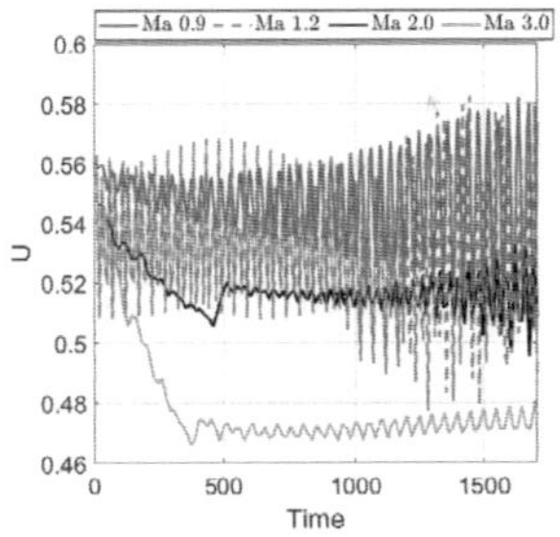

Fig. 1. Time variation of the streamwise velocity fluctuations downstream of the gap trailing-edge, at a distance δ^* from the wall, for a range of freestream Mach numbers.

[†]Felipe O. Aguirre is with the Sao Carlos School of Engineering, Department of Aeronautical Engineering, University of Sao Paulo, Av Trabalhador sao-carlense - 400, Sao Carlos/SP, Brazil, felipe.aguirre@usp.br

[◇]Paulo C. V. Paino is with the Sao Carlos School of Engineering, Department of Aeronautical Engineering, University of Sao Paulo, Av Trabalhador sao-carlense - 400, Sao Carlos/SP, Brazil, paulopaino@usp.br

[‡]Hadar Ben-Gida is with the Faculty of Aerospace Engineering, Technion Israel Institute of Technology, Haifa 3200003, Israel, bengida@technion.ac.il

[⋆]Marcello A. F. Medeiros is with the Sao Carlos School of Engineering, Department of Aeronautical Engineering, University of Sao Paulo, Av Trabalhador sao-carlense - 400, Sao Carlos/SP, Brazil, marcello@sc.usp.br

[1]https://github.com/marlonsmathias/GATT_DNS

References

[1] S. Beguet, J. Perraud, M. Forte, and J.-Ph. Brazier, "Modeling of Transverse Gaps Effects on Boundary-Layer Transition," *Journal of Aircraft*, vol. 54, no. 2, pp. 794–801, 2017.

[2] J. Crouch, V. Kosorygin, M. Sutanto, and G. Miller, "Characterizing surface-gap effects on boundary-layer transition dominated by Tollmien–Schlichting instability," *Flow*, vol. 2, no. E8, 2022.

[3] J. E. Rossiter, "Wind-tunnel Experiments on the Flow Over Rectangular Cavities at Subsonic and Transonic Speeds," Ministry of Aviation; Royal Aircraft Establishment; RAE Farnborough, Tech. Rep., 1964.

[4] G. A. Brès and T. Colonius, "Three-dimensional instabilities in compressible flow over open cavities," *Journal of Fluid Mechanics*, vol. 599, pp. 309–339, Mar. 2008.

[5] R. Mack, "The stability of compressible flow over a flat plate," *Journal of Fluid Mechanics*, vol. 21, pp. 513-529, 1964.

Manuscript 70 submitted to 64th Israel Annual Conference on Aerospace Sciences (IACAS). Received January 9, 2025.

Stability analysis of shear flows and boundary layers via novel stability criterion that utilizes the small gain theorem

Ofek Frank-Shapir*, Igal Gluzman†

Technion - Israel Institute of Technology, Haifa 3200003, Israel

In linear stability theory (LST), the critical Reynolds number is determined for infinitesimally small disturbances, which trigger the shear flow to become unstable. However, it is well known from experiments that the flow can become unstable at higher or lower Reynolds numbers than a critical Reynolds number computed via classical LST for infinitesimally small disturbances. It illustrates that disturbance magnitude has an impact on the stability of flows, and thus, the stability criterion should account for finite disturbance magnitude.

We use concepts from the fields of robust and nonlinear control theory for developing a novel approach for analyzing the stability of shear flows. We propose a novel stability criterion based on the small gain theorem (Zhou et al.,1995) coupled with unstructured and structured input-output analyses. The criterion provides a bound on the magnitude of velocity perturbations that ensure stability in the Lyapunov sense. Our stability criterion is used to study the instability of three canonical base flows—plane Couette, plane Poiseuille, and Blasius. When only linear mechanisms are considered, we show that the predictions of our stability criterion converge to the results of LST, i.e., predicting a Tollmien-Schlichting (TS) mode triggering instability at the critical Reynolds number that LST predicts for plane Poiseuille, and Blasius flows. By incorporating nonlinear interactions via a structured input-output approach, our analysis provides an upper bound of a finite value for the disturbance under which the flow remains stable. Our novel stability criterion shows that all considered canonical base flows in our study can become unstable at different sub-critical Reynolds numbers for a certain magnitude of finite value of velocity perturbation, which is supported by observed transition at subcritical Reynolds numbers in experimental studies for Couette flow (e.g., Tillmark & Alfredsson 1992), plane Poiseuille flow (Nishioka and Asai 1985), and Blasius boundary layer (Elofsson & Alfredsson 2000).

References

(Elofsson & Alfredsson 2000) Elofsson, Per A & Alfredsson, P Henrik 2000 An experimental study of oblique transition in a Blasius boundary layer flow. European Journal of Mechanics-B/Fluids 19 (5), 615–636.

(Nishioka and Asai 1985) Nishioka, M. and Asai, M., 1985. Some observations of the subcritical transition in plane Poiseuille flow. Journal of Fluid Mechanics, 150, pp.441-450.(Schmid, 2007)

(Tillmark & Alfredsson 1992) Tillmark, Nils & Alfredsson, P Henrik 1992 Experiments on transition in plane Couette flow. Journal of Fluid Mechanics 235, 89–102.

(Zhou et al.,1995) Zhou, Kemin, Doyle, John C & Glover, Keith 1995 Robust and Optimal Control. Upper Saddle River, NJ: Pearson

*Graduate Student, Faculty of Aerospace Engineering. E-mail: ofekfr@campus.tecnion.ac.il
†Assistant Professor, Faculty of Aerospace Engineering. E-mail: igal.gluzman@technion.ac.il

Manuscript 30 submitted to 64th Israel Annual Conference on Aerospace Sciences (IACAS). Received December 27, 2024.

Genetic Algorithm-Based Approach to Load Distribution in Full-Scale Structural Test Design

Boris Dorfman[1], Efrat Pinhas[1], Yuval Freed*[1]

[1]*Aviation Group, Israel Aerospace Industries, Ben-Gurion International Airport, Israel*

Structural tests are a cornerstone of aircraft design, ensuring that airframes can withstand the extreme loads and operational conditions encountered during their service life. These tests, conducted at both the component and full-scale levels, validate theoretical models and computational predictions, providing critical insights into structural integrity, durability, and safety margins. Full-scale static tests, in particular, replicate the most demanding load scenarios an aircraft may experience, verifying that the structure complies with regulatory requirements and design specifications. They are essential for detecting unforeseen weaknesses, such as unexpected stress concentrations or material behavior anomalies, which could compromise performance or safety. By uncovering potential issues early, structural tests reduce the risk of in-service failures, enhance reliability, and contribute to the development of lighter, more efficient designs that meet stringent industry and safety standards.

Designing how to apply discrete loads in full-scale static tests is a critical aspect of structural evaluation in aircraft design. The goal is to replicate, as accurately as possible, the complex load distributions that an airframe will experience in real-world operational conditions. This involves

*Corresponding author, Dr. Yuval Freed E-mail: freedyuval@gmail.com

Manuscript 10 submitted to 64th Israel Annual Conference on Aerospace Sciences (IACAS). Received December 16, 2024.

determining the optimal number, magnitude, and locations of load application, which are typically introduced through hydraulic actuators, weights, or other mechanical means. Minimizing the number of discrete loading points is essential to reduce costs and accelerate testing duration (especially in full scale fatigue tests). However, obtaining the optimal number and location of discrete loading points, all without reducing accuracy, is a complex task. To this end, this study utilizes genetic algorithm framework for this task. A remarkable reduction of ~30% of loading channels was achieved for benchmark problems (taken from platforms tested in the past) using the proposed approach. Other possible applications using this approach will be discussed as well.

Machine Learning-Based Surrogate Models for Predicting Crack Growth in Aerospace-Grade Aluminum Alloys

Yuval Freed*[1]

[1]*Aviation Group, Israel Aerospace Industries, Ben-Gurion International Airport, Israel*

Fatigue in metallic structures has posed significant challenges in aviation since the mid-20th century. The adoption of metals as primary construction materials for aircraft led to fatigue cracks and catastrophic failures, such as the Comet I disaster (1954), the F-111 wing failure (1969), and the Boeing 707 crash (1977). These tragedies spurred the aviation industry to innovate fatigue assessment and prevention techniques.

By the 1970s, fatigue analysis for airframes began integrating fracture mechanics, particularly for predicting crack growth. Paris' law (1961) established a linear relationship between crack growth rates and stress intensity in a log-log scale. However, non-linearities—especially in small and long cracks—and the influence of the stress ratio (R-ratio) on growth rates were later discovered. To address these complexities, refinements were made: Forman et al. (1969) incorporated R-ratio and critical fracture toughness (Kc) into models for large cracks, while Elber's concept of crack closure (1970) improved predictions for cyclic loading. Willenborg et al. (1971) introduced load-interaction concepts, further enhancing modeling accuracy.

The adoption of damage tolerance principles in military (1974) and civilian (1978) aviation

*Corresponding author, Dr. Yuval Freed E-mail: freedyuval@gmail.com

Manuscript 11 submitted to 64th Israel Annual Conference on Aerospace Sciences (IACAS). Received December 16, 2024.

solidified fracture mechanics as a cornerstone in airframe design and inspection programs, emphasizing the need for accurate crack growth predictions.

Recent advancements in computational resources have opened new doors for data-driven methods to analyze complex phenomena like fatigue crack growth. Machine learning approaches, such as neural networks and support vector regression (SVR), have been successfully applied to materials like aluminum and titanium. Studies by Bao et al. (2021) and Zafar et al. (2022) demonstrated the efficacy of these techniques for predicting crack growth behavior. This study focuses on developing a surrogate model to predict crack growth in 7075-T6 aluminum, a key aviation alloy with a complex crack growth response. Influencing factors include the stress intensity range (ΔK), R-ratio, and specimen thickness. Notably, significant non-linearities occur in the ΔK range of 8 to 12, add complexity to predictions, as presented in Figure 1.

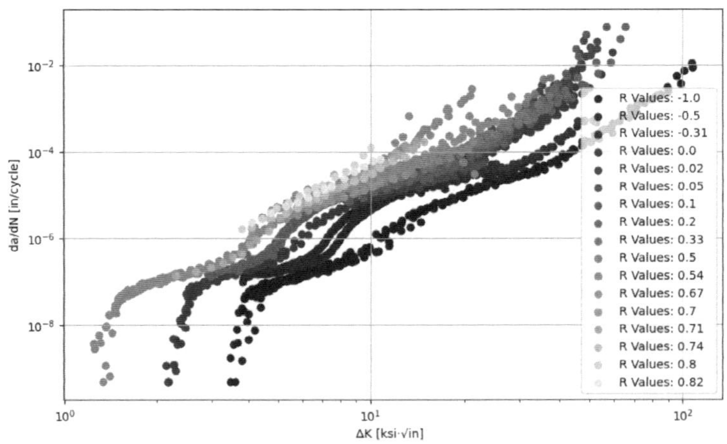

Fig. 1. Crack growth behavior for 7075-T6 with different R-ratios

.

To address these challenges, four machine learning regression algorithms were evaluated: Random Forest (RF), Gaussian Process Regression (GPR), K-Nearest Neighbors (KNN), and Kernel

Regression (KR). Their performance was compared to select the most effective model for predicting crack growth under various fatigue stress spectra. Validation tests revealed that while all algorithms produced reasonable predictions, Random Forest regression consistently outperformed the others in predictive accuracy.

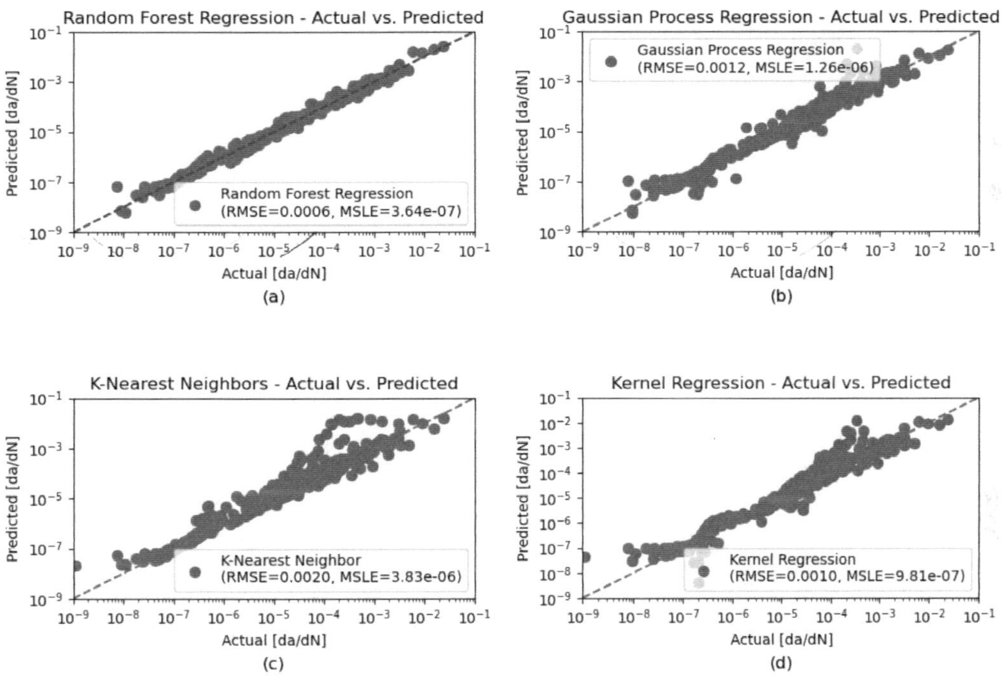

Fig. 2. Predictions vs. actual algorithm test data by means of (a) Random Forest regression, (b) Gaussian Process Regression, (c) K-Nearest Neighbors regression and (d) Kernel regression

References:

Bao, H., Wu, S., Wu, Z., Kang, G., Peng, X. and Withers, P.J., "A machine-learning fatigue life prediction approach of additively manufactured metals", Engineering Fracture Mechanics, 242, p.107508, 2021 (DOI: https://doi.org/10.1016/j.engfracmech.2020.107508)

Elber, W., "Fatigue crack closure under cyclic tension", Engineering Fracture Mechanics, 2, p. 37-45, 1970 (DOI: https://doi.org/10.1016/0013-7944(70)90028-7)

Forman, R.G., Kearney, V.E and Engle, R.M., "Numerical analysis of crack propagation in cyclic-loaded structure", Journal of Basic Engineering, 89, p. 459-464, 1967 (DOI: https://doi.org/10.1115/1.3609637)

Paris, P.C., Gomez, M. and Anderson, W.E., "A rational analytic theory of fatigue", Trends in Engineering, 13, University of Washington, USA, 1961

Willenborg, J., Engle, R.M. and Wood, H.A., "A crack growth retardation model using effective stress concept", AFFDL-TM-71-FBR, Wright Patterson Air Force Laboratory, 1971

Zafar, M.H., Younis, H.B., Mansoor, M., Moosavi, S.K.R., Khan, N.M. and Akhtar, N. "Training deep neural networks with novel metaheuristic algorithms for fatigue crack growth prediction in aluminum aircraft alloys", Materials, 15(18), p.6198, 2022 (DOI: https://doi.org/10.3390/ma15186198)

Leveraging Foundation Model approach in Fluids Mechanics Systems engineering

*Shaul Eliahou-Niv, +Yotam Gardosh, +Asaf Shiloah

This paper aims to explore how current Artificial Intelligence (AI) technology contributes to the enhancement and acceleration of systems engineering tasks. While systems engineering has broader roots, aeronautics has been a pivotal area for its application and evolution. The integration of Artificial Intelligence (AI) in systems engineering marks a significant evolutionary step in managing complex projects. AI can optimize system designs and processes, leading to cost savings and efficiency improvements. Foundation Models can perform a wide range of disparate tasks with a high degree of accuracy based on input prompts. Unfortunately, despite all the capabilities listed above, Large language models (LLM's) have significant limitations in field of fluid mechanics. The limitation of LLMs in dealing with tasks in fluid dynamics stems from their lack of specialized knowledge and inability to perform complex numerical simulations required for accurate predictions in this field. Our conjecture is that these limitations arise because these generative models have not been trained or refined with required information and data on fluid dynamics phenomena. In our research we employed several strategies to enhance the accuracy of LLMs, including Prompt Engineering, Retrieval-Augmented Generation (RAG) and Error Detection and Correction. RAG is an AI framework that combines the strengths of traditional information retrieval systems with the capabilities of LLM's. Actually, the RAG system is an information retrieval approach designed to overcome the limitations of using a LLM directly. By combining unique and designated data of the user and the world knowledge with LLM language skills grounded generation is more accurate, up-to-date, and relevant to the user specific needs. Due to the compute cost, data preparation time and the required resources, using RAG without training or fine-tuning is an attractive proposition. However, challenges arise when using large language models for information extraction such as performance with long text. In Prompt tuning one adjusts the content of the prompt that is passed to the model to guide the model to generate output that matches a pattern that was specified. The basic foundation model and its parameters are not edited. Only the prompt input is altered. Prompt-tune model, the underlying foundation model can be used to address different engineering needs without being retrained each time. This reduces the computational needs and inference costs. By using both methods and our unique corpus in the RAG Framework we were able to reduce the error rate of different LLM's by 9.07% - 33.33% depending on the model. In our research we outline a structured approach to building a domain-specific RAG database, applying techniques such as text splitting, cleaning and relevance checking to enhance retrieval effectiveness. Key sections cover the construction and optimization of the RAG system, followed by an evaluation using the LM-Harness framework to assess model accuracy and confidence on system engineering benchmarks. Initial results provide insights into the model's capabilities and limitations, while future work will expand the database, improve relevance filtering, and test larger models to further understand LLM applicability in systems engineering contexts. We hope that this technical report will serve as a starting point for further collaboration and efficient knowledge sharing between reserchers from the fluid mechanics community to bridge the existing gap between the pre-trained models and the expert required information for the accurate rag infrastructure. This can accelerate the onboarding process for quickly collect and distill relevant information, documentation and best practices for accurate querying of the LLMs. This document represents a mid-stage report, setting the foundation for ongoing and future developments in the field.

*Director, Applied Research IAI. Senior Research Fellow, The faculty of Aeronautics Technion
+Undergraduate student, Faculty of Computer Science, HUJI.

This research was made possible thanks to the generous support of the Bernard M. Gordon Center for Systems Engineering at the Technion.

Inner Outer Predictive Model Applied to Atmospheric Surface Layer Turbulence

Maayan Shimoni, Ian Jacobi, Anna Clarke
Faculty of Aerospace Engineering
Technion - Israel Institute of Technology

January 7, 2025

Abstract

We adapt the widely-used inner-outer predictive model for wall-bounded turbulence to predict wind turbulence in the atmospheric surface layer (ASL) under varying thermal stability conditions. Marusic, Mathis and Hutchins (2011) first proposed a predictive inner-outer model, that is, the Marusic-Mathis-Hutchins (MMH) model, for the prediction of the streamwise velocity time series in the inner region of wall turbulence based on a large-scale velocity signature measured at an outer position in the logarithmic region of the flow. The model uses universal time-series which are modified by the outer large scale signature through superposition and modulation, expressed by three constants that are empirically determined. Data from the CASES-99 experiment is used to extract the model components and test its effectiveness to atmospheric, complex flow of various stability conditions. Through the investigation of the MMH model superposition and modulation constants- inclination angle, peak correlation, and modulation coefficient, we have found that the inclination angle increases with thermal instability, due to the upward lifting effect on coherent structures. Conversely, the modulation coefficient decreases with increasing instability as the spatial attributes of the flow structures change from streamwise to vertically dominated. A key contribution of this study is the investigation of the demodulated 'universal' signal, obtained in the calibration process, by normalizing the turbulent signals by the friction velocity u_τ, and then applying the demodulation using the three constants. Our analysis reveals that the normalization by u_τ is insufficient for the atmospheric data, as it fails to accurately capture the signal's variance, as even for the same stability condition and approximately same friction velocity, the turbulent signals can still differ in their variances. This inadequacy propagates to the demodulated signals, resulting in significant discrepancies in the variance and power spectrum, thereby challenging their universality. To address this issue, we introduce normalization by the outer signal variance, which significantly enhances the demodulation process. This adjustment ensures that the demodulated signals are much closer to being 'universal' across various stability conditions, supporting the concept of a universal inner signal that is modified through modulation and superposition of the large scale outer motions. Using the characterization of the MMH constants with the thermal stability, we prove that the MMH model can be applied to atmospheric, thermally affected data, using a modified version which includes additional information about the outer flow variance.

Manuscript 49 submitted to 64th Israel Annual Conference on Aerospace Sciences (IACAS). Received January 7, 2025.

Flight Course Maneuver Optimization

Ido Braun

Abstract

In the context of a fighter jet operating within a missile-protected environment and tasked with deploying munitions against a designated target, the goal of the problem is to optimize the aircraft's survivability. The aircraft faces continuous threats from surface-to-air missiles (SAM) during its mission, necessitating a strategic approach to minimize its exposure in the hostile area. To achieve this, the pilot aims to release the munition from the maximum possible range, requiring a higher altitude due to the altitude-limited range performance of the munition. However, operating at higher altitudes compromises the aircraft's performance and evasion maneuverability, posing a risk to prolonged exposure in the threatened area. The objective of this research is to identify the optimal altitude and maneuver course that maximizes the aircraft's survivability.

All basic assumptions of the research are made aiming to provide a realistic approach to solving the problem. The first assumption is that a competent pilot should be able to receive the results and use them as instructions to fly the aircraft throughout the mission. Hence, the control commands of the aircraft's model are the load factor, roll rate and thrust, all of which the pilot can control directly using stick and throttle commands. In that regard, the pilot commands must be pilotable and non-complex. The second assumption is that the pilot is unaware of the missile's whereabouts except for the missile launcher location, and so the aircraft may not be able to actively maneuver in order to evade the missile in closed-loop. Therefore, it is desired to find the open-loop maneuver which provides optimal survivability.

The presented research regards computing a full dynamic 3DOF model for the fighter jet aircraft, interceptor missle, and the range envelope of the munition. A time optimal maneuver is calculated in order to climb to munition toss altitude from the furthest possible range, and afterwards an escape maneuver is performed without the aircraft's acknowledgement of the pursuing missile. The missile is launched with the purpose of intercepting the aircraft, and its miss distance is computed. The problem will be simulated in various conditions to identify the optimal toss altitude and evasion maneuver for survivability, which is defined by the miss distance of the missile's interception attempt. The conditions are the munition toss altitude and velocity, initial aircraft azimuth relative to the target, and aircraft escape maneuver turn angle. A bi-level optimization problem is formulated whereby the inner level is the time optimal maneuver and the outer level is a 4-dimensional (black box) optimization problem, which is solved using iterative methods such as Golden Section Search, Nelder-Meade, Monte-Carlo, Bayesian Optimization and Cross-Entropy. In addition, algorithms have been developed in order to reduce the optimal aircraft maneuver into several pilotable commands so that a pilot could understand and pilot these instructions. Algorithms were created both for the climb maneuver, and the escape maneuver resembling Split-S. Other cases have been investigated in which there are several missiles aiming to intercept the aircraft, or in which there is an uncertainty regarding the missile's launcher location.

Manuscript 79 submitted to 64th Israel Annual Conference on Aerospace Sciences (IACAS). Received January 12, 2025.

Hybrid Belief Space Planning with Coupled Semantic-geometric Models

Tuvy Lemberg and Vadim Indelman *

January 9, 2025

Robots often use geometric-semantic representations of the environment to perform complex tasks safely. The semantic map of the environment can influence robot actions and constraints. For an autonomous vehicle, different traffic signs located in different locations will cause a variety of constraints. In general, semantic observations of the environment depend on both semantic and geometric variables. A traffic sign that is observed from the front will be correctly identified and will indicate that it is in front of the sign. However, looking at a traffic sign from behind can result in wrong classification and it will indicate to the robot that it is behind the sign. Most existing studies model semantic observations independently of viewpoint, resulting in statistical information loss. Several recent studies considered coupling, but they pruned semantic hypotheses to reduce computational complexity. As part of our work, we present a method for estimating the value function while preserving all semantic hypotheses.

*The authors are with the Department of Aerospace Engineering, Technion - Israel Institute of Technology, Haifa 32000, Israel. tuvy@campus.technion.ac.il, vadim.indelman@technion.ac.il.

Manuscript 59 submitted to 64th Israel Annual Conference on Aerospace Sciences (IACAS). Received January 9, 2025.

Real-time Sky Object Detection and Classification using YOLO Algorithm

As the population of satellites in Low Earth Orbit (LEO) continues to grow, the need for advanced monitoring systems has become increasingly critical. This expansion underscores the importance of real-time, automated solutions to ensure the sustainable management of space activities.

This research investigates a novel application of the You Only Look Once (YOLO) algorithm for real-time detection and classification of sky objects. YOLO, a state-of-the-art convolutional neural network (CNN), is renowned for its capability to deliver rapid and accurate object recognition through single-pass predictions. Using imagery captured by two strategically positioned all-sky cameras, approximately 180 km apart, this study successfully identifies and categorizes objects such as satellites, space debris, and meteors with high precision.

The system's performance is driven by YOLO's efficient feature extraction capabilities, which process entire images and generate bounding boxes around detected objects in a single operation. This approach significantly outperforms traditional methods, offering superior speed without sacrificing accuracy. The all-sky cameras, selected for their cost-effectiveness and wide-field observational range, enable the detection of objects with magnitudes up to +4, making them well-suited for monitoring LEO satellites and debris.

Model training was conducted using a dataset of approximately 1,000 curated all-sky images, optimized to enhance detection performance across diverse scenarios. Validation was carried out using a confusion matrix to evaluate classification accuracy across multiple object categories, yielding promising results. Additional training data is expected to further improve the system's robustness and reliability.

By integrating YOLO with cost-effective imaging technology, this study offers a scalable and efficient framework for real-time detection and classification of sky objects. It addresses the growing challenges posed by the expanding satellite population and increasing space debris in LEO. Furthermore, this work provides a replicable framework applicable to other types of space threats, emphasizing its strategic importance in enhancing national airspace security through advanced real-time monitoring capabilities.

Manuscript 61 submitted to 64th Israel Annual Conference on Aerospace Sciences (IACAS). Received January 9, 2025.

Detecting GPS Spoofing Incidents Using Variational Autoencoders

Dr. Barak Or

MetaOr Artificial Intelligence, Haifa, Israel

Google and Reichman Tech School, Herzliya, Israel

-Abstract-

The dependence on GPS for navigation in autonomous vehicles (AVs) introduces significant vulnerabilities, particularly the risk of GPS spoofing - a cyberattack where false signals mimic legitimate GPS data to mislead navigation systems. This study addresses the challenge of detecting such incidents using a robust Variational Autoencoder (VAE)-based anomaly detection framework, trained on normal GPS data to identify deviations indicative of spoofing.

As AVs rely heavily on accurate GPS data for real-time positioning, collision avoidance, and emergency response, ensuring the integrity of these signals is paramount. Traditional countermeasures, such as encryption and redundant systems, often fail against advanced spoofing techniques. Therefore, we developed an advanced detection system leveraging the representational power of VAEs to monitor and analyze GPS navigation patterns.

The suggested approach involved collecting and processing normal GPS signal data from various real-world driving scenarios, encompassing diverse vehicles, locations, and conditions, concluding over four hours of driving data. Spoofed GPS signals were synthetically generated to simulate realistic attack scenarios. The VAE model, trained on this dataset, was designed to learn the latent space representations of normal GPS patterns. During testing, the VAE detected anomalies by comparing real-time data against the learned patterns, with deviations flagged as potential spoofing incidents.

The VAE model achieved an accuracy of 97.9% for anomaly detection. The model correctly identified 704 spoofed signals, with only 30 misclassifications. This high level of accuracy underscores the model's capability to distinguish between authentic and spoofed GPS data effectively.

The proposed system's application extends beyond terrestrial vehicles to aerial systems, such as drones and aircraft, where GPS integrity is critical. For example, incidents like the Iran–U.S. RQ-170 drone interception highlight the importance of detecting and mitigating GPS spoofing in aerospace domains. This research lays the groundwork for integrating such systems into real-time navigation technologies, enhancing the safety and reliability of modern transportation systems.

This study presents a scalable, effective solution for GPS spoofing detection using VAEs, demonstrating its potential to safeguard autonomous and aerial navigation systems against emerging cyber threats.

Manuscript 80 submitted to 64th Israel Annual Conference on Aerospace Sciences (IACAS). Received January 12, 2025.

Enhancing Predictive Maintenance with Transformer-based Deep Neural Network

Dr. Barak Or

MetaOr Artificial Intelligence, Haifa, Israel

Google and Reichman Tech School, Herzliya, Israel

-Abstract-

Accurately predicting the Remaining Useful Life (RUL) of engines is essential for ensuring flight safety, minimizing downtime, and optimizing maintenance costs. Traditional model-based approaches and classical statistical methods, while widely used, often fail to address the complexities of modern aero engines, particularly when faced with incomplete or partially available sensor data. Missing or noisy inputs, caused by hardware malfunctions, data transmission errors, or adverse operating conditions, pose significant challenges for these methods, leading to suboptimal predictions and increased risks of unexpected failures or inefficient maintenance schedules.

Transformers, originally developed for natural language processing, provide a promising solution to these challenges. With their ability to model long-range dependencies and adapt to missing or partial data, transformer architectures excel in handling sequential data, making them particularly suited for predictive maintenance tasks in aerospace. By processing historical operational data and real-time sensory inputs, transformers improve RUL predictions and real-time failure forecasting, even in scenarios where some sensor data is unavailable. Their attention mechanisms prioritize available, high-confidence data while mitigating the impact of missing or noisy inputs.

Unlike classical statistical models, which often simplify dependencies and fail to account for nuanced degradation patterns, transformers leverage advanced masking and imputation strategies to simulate missing data during training, enhancing their ability to generalize to real-world scenarios. Furthermore, transformers' adaptability makes them a powerful tool for integrating predictive maintenance systems with onboard robotics and automated diagnostic systems, facilitating proactive interventions such as operational adjustments or maintenance scheduling.

By addressing the limitations of traditional methods, this approach underscores the great potential of deep learning, and in particular transformers model, in predictive maintenance.

Manuscript 86 submitted to 64th Israel Annual Conference on Aerospace Sciences (IACAS). Received January 14, 2025.

Remotely Piloted Aircrafts Automatic Takeoff and Landing Performance Evaluation

Author: Tsoof Joels (ESL)

Remotely Piloted Aircrafts (RPA), also known as Unmanned Aerial Vehicles (UAV), have been used in various applications, both military and civil. With an expected annual growth of 5%, the level of automation is rising to meet the need for certified personnel. Traditionally, RPAs are operated during takeoff and landing (TOL) by an External Pilot (EP). In recent years, automatic takeoff and landing (ATOL) systems have been implemented in RPA systems to reduce dependency on EPs, who require specific skills and extensive, expensive training. Nowadays, several RPA systems, including medium (up to 1600 kg) and large (a couple of tons) weight categories, do not have the option to be operated by an EP.

Flight testing is a mandatory part of RPA development, during which various abilities and performances, including ATOL, need to be evaluated. During development, EPs play a crucial role in ATOL testing, having the ability to judge RPA performance and abort the automatic procedure if there are any safety concerns. Thus, systems designed without EP controls pose a design challenge to temporarily include EP controls and an operational challenge to the EP, whose commands may be interpreted as disturbances by the autopilot.

This study proposes a method to evaluate RPA ATOL performances during the flight test period, which does not depend on certified EPs. Previous work used a classification algorithm for takeoffs and landings based on input from flight crew members, including the internal pilot (IP) and the EP. The classification algorithm was recently upgraded to use the electro-optic payload (EO) available on board. During ATOL, the EO is locked on a reference target, and statistical analysis of angular data in relation to the classification algorithm was used to consider replacing EP input with EO input.

The presentation will include details and results from flight tests of ATOL classifications using both the primary and the upgraded algorithm.

Manuscript 23 submitted to 64th Israel Annual Conference on Aerospace Sciences (IACAS). Received December 23, 2024.

Neural Inertial Dead Reckoning and Fusion

Itzik Klein*
The Autonomous Navigation and Sensor Fusion Lab, Hatter Department of Marine Technologies, University of Haifa, Israel

The use of quadrotors in military and civilian applications has increased significantly in the last decade. Those include construction, deliveries, surveillance, industry, marine science, mapping, emergency response, and law enforcement [1]. For quadrotors to accomplish the challenging tasks assigned to them, an accurate navigation system is essential. The fusion of a global navigation satellite system (GNSS) receiver and an inertial navigation system (INS) is one of the most common ways quadrotors navigate outside. This type of fusion is usually performed using a model-based extended Kalman filter framework. Sometimes, the quadrotor's navigation solution is dependent solely on its inertial navigation system. Consequently, the navigation solution drifts in time due to errors and noises in inertial sensor readings. To mitigate the inertial drift we offered QuadNet [2], a hybrid framework for quadrotor dead reckoning to estimate the quadrotor's three-dimensional position vector at any user-defined time rate.

In this work, we propose approaches to extend deep-learning assisted inertial sensing and fusion capabilities during periodic motion. We offer an empowered network architecture, mini-QuadNet (MQN), to accurately regress the change in distance of the platform. Using this network, we drive a hybrid approach for a neural-inertial fusion filter (with and without GNSS updates) as shown in Figure 1.

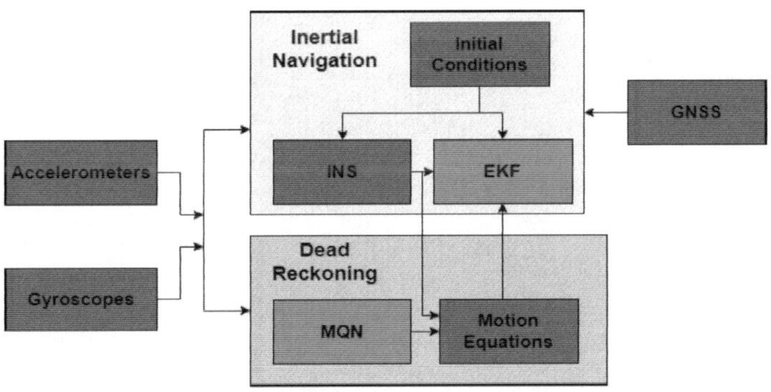

Figure 1. Neural inertial fusion block diagram. MQN-driven position and GNSS updates, when available, are used as external updates to the navigation filter.

A dataset of 337 min of data collected from four sets of inertial sensors mounted on a mobile robot and a quadrotor was used to evaluate our approaches. Figure 2 provides the position RMSE of the INS/GNSS fusion (baseline) and INS/GNSS/MQN (ours) on the test dataset for all four IMUs. In the quadrotor dataset, the addition of the MQN updates reduced the RMSE from 5.6 to 4.3 m, leading to a 21.5% improvement, and in the mobile robot, the error decreased from 5.9 to 4.2 m, resulting in an improvement of 28.8%.

* Email: kitzik@univ.haifa.ac.il

Manuscript 16 submitted to 64th Israel Annual Conference on Aerospace Sciences (IACAS). Received December 21, 2024.

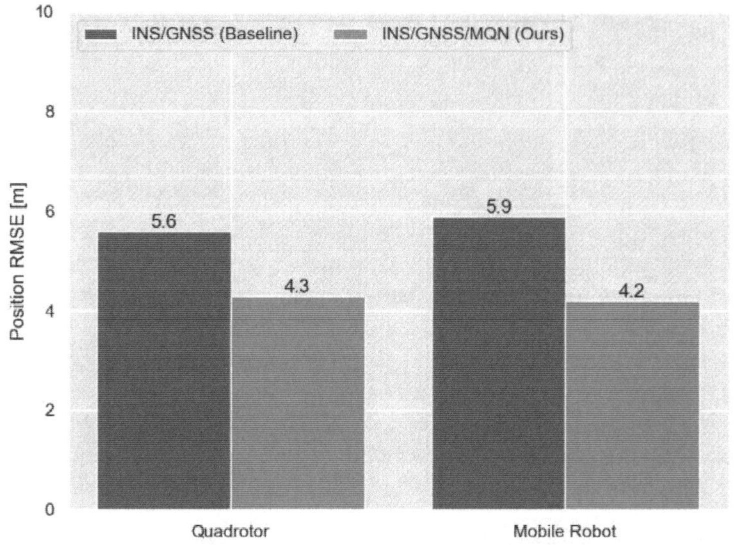

Figure 2. Position RMSE without (baseline) and with (ours) MQN 1-Hz updates.

References

[1] Idrissi, M., Salami, M., and Annaz, F., "A Review of Quadrotor Unmanned Aerial Vehicles: Applications, Architectural Design and Control Algorithms," Journal of Intelligent and Robotic Systems, vol. 104, no. 2, pp. 1-33, 2022.

[2] Shurin, A. and Klein, I., "QDR: A Quadrotor Dead Reckoning Framework," IEEE Access, vol. 8, pp. 204433–204440, 2020.

Clustering of particle-laden flows in synthetic turbulence

Boaz Ofarim*, Orr Avni†, Yuval Dagan‡

December 31, 2024

The dispersion and clustering of particles in particle-laden turbulent flows represent a crucial challenge in fluid mechanics, yet our understanding of such phenomena is far from complete. Particle interactions in turbulent flows are frequently encountered in both natural environments and engineering areas, such as rain formation, dispersion of pollutants in the atmosphere, blood flows, droplets in combustion chambers, aluminum particles in solid rocket motors, and more. In a particle-laden flow, a dispersed medium, such as dust or droplets, is distributed throughout a carrier flow, either a gas or a liquid.

Due to the considerable difficulties in computationally simulating particle-laden turbulent flows [3], this research employs the Kinematic Simulations (KS) approach, chosen for its computational efficiency and relative accuracy. In KS, a series of Fourier modes replicates a flow field, satisfying continuity and turbulence-like energy spectra [4]. KS has been utilized in various research contexts: Fung and Vassilicos [5], as well as Chen, Goto, et al. [2] employed KS to explore particle pair-dispersion and clustering in turbulence, respectively.

Until now, there has been limited research integrating KS flow fields with thermodynamic processes such as evaporation, condensation, and nucleation. Avni and Dagan [1] investigated droplet dynamics in Burgers vortices, incorporating evaporation and condensation phenomena. This study adopts the same thermodynamic model while coupling it with the KS flow field. Such thermodynamic effects can play a significant role in droplet clustering, a factor that has not been previously addressed.

References

[1] AVNI, O., AND DAGAN, Y. Droplet dynamics in burgers vortices. I. mass transport. *Physical Review Fluids 8*, 8 (2023), 083604.

[2] CHEN, L., GOTO, S., AND VASSILICOS, J. Turbulent clustering of stagnation points and inertial particles. *Journal of Fluid Mechanics 553* (2006), 143–154.

[3] ELGHOBASHI, S. Direct numerical simulation of turbulent flows laden with droplets or bubbles. *Annual Review of Fluid Mechanics 51*, 1 (2019), 217–244.

[4] FUNG, J. C. H., HUNT, J. C. R., MALIK, N. A., AND PERKINS, R. Kinematic simulation of homogeneous turbulence by unsteady random fourier modes. *Journal of Fluid Mechanics 236* (1992), 281–318.

[5] FUNG, J. C. H., AND VASSILICOS, J. C. Two-particle dispersion in turbulentlike flows. *Physical Review E 57*, 2 (1998), 1677.

*Faculty of Aerospace Engineering, Technion.
†School of engineering, Brown University.
‡Faculty of Aerospace Engineering, Technion.

Manuscript 36 submitted to 64th Israel Annual Conference on Aerospace Sciences (IACAS). Received January 1, 2025.

Title: On the Noise Modulation of Small-Scale Low-Reynolds Number Rotors of Different Materials

Authors: Aharon Karon (IAI), Aleksandra Kvurt (IAI), Jenya Kazarin (Technion), and Hadar Ben-Gida (Technion)

Abstract:

Small-scale rotors operating at low Reynolds numbers, commonly used on small unmanned aerial vehicles (UAVs), can generate noise with high-frequency modulation. This noise results from the coupling of blade passing frequency (BPF) and high-frequency content, and it is particularly significant in small-scale rotor designs. The specific modulation effect is the amplitude of the high-frequency noise by the blade-pass frequency (BPF). This phenomenon can affect the human perception of the noise due to the time-varying nature of the amplitude, thus impacting human annoyance and early detection in defense applications. Characterizing this phenomenon will assist in the design of rotors and make the whole UAV system less annoying and detectable. In this study, rotors of different geometries and materials were studied. The commercial off-the-shelf (COTS) DJI Phantom 3 9450 rotor and APC 11x10E and 9x6E rotors were selected as they are widely used in research. The geometries of these rotors were scanned, and additional rotors of these geometries were fabricated from aluminum and carbon fiber, as well as 3D printed from nylon. Performance (forces and moments) and acoustic measurements were acquired for the above rotors at hover conditions and for a range of RPMs in the anechoic wind tunnel in the Faculty of Aerospace Engineering at the Technion. Both bi-coherence and wavelet analysis techniques were used to characterize the modulation effects. The results of this study improve the understanding of the noise modulation phenomenon in small-scale low-Reynolds rotors operating in hover, as well as its effect on human perception. The results also reveal how noise modulation varies with different rotor materials. The outcome of this research will assist in rotor design and limit noise impact and detection by human observers.

Manuscript 29 submitted to 64th Israel Annual Conference on Aerospace Sciences (IACAS). Received December 25, 2024.

Utilization of computer vision algorithms for the characterization of coupled interactions between bubbly shocks and cavitation cloud

Elad Zur* and Igal Gluzman†
Technion – Israel Institute of Technology, Haifa 3200003, Israel

I. Abstract

The most severe cavitation noise and erosion damage in internal flows are produced when many bubbles in the cloud collapse close to each other near a solid surface, resulting in an inward propagating shock wave. Cloud cavitation exhibits extremely violent behavior at high void fractions in cavitating flows due to the enhancement of the pressure waves when bubbles collapse. The bubbly shocks that are emitted from the cavitation cloud collapse travel through the bubbly domain, leading to complicated interactions between the bubbly shock and the cavity voids, which present significant challenges for experimental measurements. It requires both an adjusted experimental setup with optical access and data processing capabilities that allow simultaneous monitoring of the spatial-temporal evolution of bubbly shock waves within the cavitation cloud alongside bubble dynamics. This study is focused on fulfilling these two requirements. For this task, we designed and constructed an experimental setup to study cavitating flows in a generic converging-diverging nozzle system, allowing high-speed imaging and high-frequency pressure measurements of cavitation flows in various liquids, including water, dodecane, and aviation fuels. The facility design is based on the setup that is used in [1] and [2]. In addition, we have reproduced and expanded the Computer Vision (CV)—based data processing code of [1] for bubble detection with new shock-wave detection capabilities that utilized the enhanced gradient shadowgraphy technique proposed by [2]. We demonstrate the new code capability to accurately capture the spatio-temporal evolution of the leading and trailing edges of bubbly shock waves in aerated cavitating flows in CD-nozzle, as well as effectively track their centers of mass. This in turn, allows us to quantify bubbly shock morphology due to their interaction with cavitation shed vortices. These developments provide valuable insights into bubbly shock behavior for modeling bubbly shocks and cavitation physics in internal flows, which is vital for improving the design and reliability of internal liquid flow systems.

References

1. Gluzman, I. and Thomas, F. O., "Characterization of bubble dynamics in the nozzle flow of aviation fuels via computer vision tools," *International Journal of Multiphase Flow*, Vol. 154, 2022, pp. 104133.

*Dept. of Aerospace Engineering. eladzur@campus.technion.ac.il
†Asst. Prof., Dept. of Aerospace Engineering. igal.gluzman@technion.ac.il

Manuscript 33 submitted to 64th Israel Annual Conference on Aerospace Sciences (IACAS). Received January 1, 2025.

2. Gluzman, I. and Thomas, F. O., "Image-based characterization of the bubbly shock wave generation and evolution in aviation fuel cavitation," *Physical Review Fluids*, Vol. 7, No. 8, 2022, pp. 084305.

Modelling steady features of cavitation in radial flow between two overlying disks with varying gaps

Samruddhi Salunke* and Igal Gluzman†
Technion - Israel Institute of Technology Israel, Haifa 32000, Israel

Cavitation in thin-gap internal radial flows occurs in many industrial components like bearings, fuel pumps, and hydraulic systems. These complex internal flow geometries include narrow flow paths and sharp turns, which give rise to large decreases in static pressure that can lead to gaseous and vaporous cavitation. Understanding the underlying cavitation physics in these thin-layer radial flow geometries is mainly motivated by the need to prevent premature material degradation and failure in these components. Recently, in [1], detailed experimental characterization and modeling of aviation fuel (JP-5) cavitation physics in radial flow in a thin layer between two stationary disks has been reported. This so-called port plate geometry is relevant to aircraft fuel pump components. In their study, the authors utilized the spatial Rayleigh–Plesset equation and barotropic model that was modified for radial flow geometry in order to predict the radial location of cavitation bubble collapse and reproduce radial pressure profiles obtained from the experiment. The model predictions of the radial location of bubble collapse and the radial pressure profiles were shown to be in excellent agreement with the experiments. However, in their study, they only considered a fixed gap between the plates. The present study extends the study of [1], where we have generalized the modeling framework for predicting the main steady features of cavitation and bubble dynamics phenomena in radial flow geometries while accounting for different gap thicknesses between two stationary disks. The model is used to characterize the variation of bubble radius, radial collapse distance, and mean radial pressure profile for JP-5 aviation fuel and distilled water across different gap heights (ranging from 0.3 to 1 mm) and different inlet-outlet pressure ratios. Model predictions for gap thickness 0.3 mm are validated against experimental results from [1]. It was found that increasing the gap between the plates results in a linear increase of the collapse distance from the disk center for all pressure ratios. The slope of the linear trend increases at higher pressure ratios. On the contrary, the radial pressure profiles in our model are scarcely affected by the gap variation. In our future study, we aim to validate the model predictions for the considered gaps in the model via experiments, further extend the model for unsteady features of cavitation and bubble dynamics phenomena in radial flow geometries, and incorporate the multicomponent characteristics of the liquids. Our model can be integrated to predict and understand the effect of cavitation on fuel pump life and performance.

References

[1] Gluzman, I., Pelster, A., Waldrop, M., and Thomas, F. O., "On cavitation in the radial flow of a thin lubricating film between two overlying disks," *Physics of Fluids*, Vol. 35, No. 2, doi:10.1063/5.0132050.

*Graduate Student, Dept. of Aerospace Engineering, samruddhis@campus.technion.ac.il.
†Assistant Professor, Dept. of Aerospace Engineering, igal.gluzman@technion.ac.il.

Manuscript 34 submitted to 64th Israel Annual Conference on Aerospace Sciences (IACAS). Received January 1, 2025.

The 64th Israel Annual Conference on Aerospace Sciences

Rapid Depressurization-Induced Flash Boiling: A Theoretical Model for Positive and Negative Pressure Ranges

Omry Magen[1], Yoram Kozak[1], Laura Di Lucchio[2], Marco Marengo[2], Tali Bar-Kohany[1,3]

[1]*School of Mechanical Engineering, Tel Aviv University, Tel Aviv, Israel*
[2] *Dep. of Civil Engineering and Architecture, University of Pavia, Pavia, Italy*
[3]*Nuclear Research Center Negev, Israel*

Flash boiling, driven by the rapid depressurization of liquids, is a critical phenomenon for optimizing the performance of aerospace fuel injection systems and propulsion components. This process enhances droplet formation and promotes finer atomization, which are critical for the development of new upper orbit thrusters using cryogenic propellants, where understanding the dynamics of oxidizer and fuel injection at near-vacuum conditions before ignition is essential [1]. This study investigates the thermodynamic principles underpinning flash boiling under rapid depressurization scenarios. Using methodologies from prior research [2, 3, 4], a model is proposed to predict the degree of metastability of a liquid during these processes, supported by published experimental results [5, 3, 6, 7].

Under homogeneous conditions, where impurities (i.e., heterogeneities) are absent, a liquid can depressurize from an initial subcooled state, crossing the equilibrium curve to reach a thermodynamic stability limit. This limit is defined by the point where the derivative of pressure with respect to volume at constant temperature is zero [8]. At this point, the liquid spontaneously changes phase, returning the system to equilibrium. However, real systems exhibit heterogeneities that trigger phase change before this limit is reached. Lienhard and Alamgir [2] introduced the heterogeneity factor, a method to quantify the significance of such heterogeneities. They found a strong correlation of this factor to the time scale of the process, i.e the depressurization rate. Building on this concept, the current study normalizes experimental results relative to idealized conditions defined by the thermodynamic limit. It is shown that the deviation from homogeneous conditions is directly proportional to the experimentally achieved depressurization rate relative to the theoretical rate required to reach the limit. This relationship is expressed as:

$$\frac{\mathrm{Ja}}{\mathrm{Ja}_{\mathrm{hom}}} \propto \frac{\Sigma}{\Sigma_{\mathrm{min}}} \tag{1}$$

where Ja and $\mathrm{Ja}_{\mathrm{hom}}$ are the experimentally acquired and homogeneous Jakob numbers, respectively. Σ represents the experimentally acquired depressurization rate, and Σ_{min} is the minimum depressurization rate required to reach the thermodynamic limit. Developing an accurate model for Σ_{min} is critical to realizing this proportionality.

By making assumptions about the thermodynamic nature of the depressurization process and the state of the liquid at the limit, a linear relationship emerges between these proportions. This relationship holds true across experiments conducted on water [5, 3, 6] and carbon dioxide [7], as shown in Fig. 1. Notably, the results demonstrate consistency not only across studies but also between different fluids. These predictive characteristics are crucial for aerospace propulsion, where accurate modeling of cryogenic propellant injection dynamics supports the design and optimization of thrusters operating under near-vacuum conditions.

Manuscript 18 submitted to 64th Israel Annual Conference on Aerospace Sciences (IACAS). Received December 22, 2024.

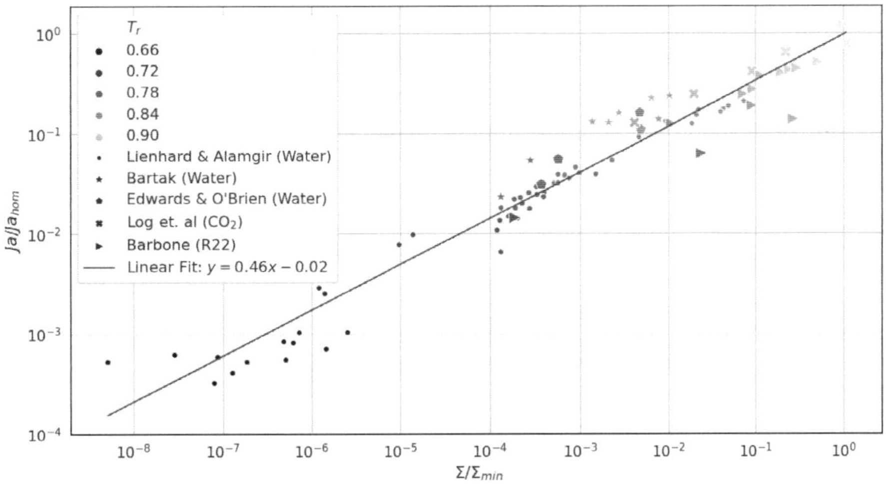

Figure 1: Linear relationship between the relative depressurization rate and the relative Jakob number.

References

[1] J. W. Gaertner, A. Kronenburg, A. Rees, J. Sender, M. Oschwald, and G. Lamanna, "Numerical and experimental analysis of flashing cryogenic nitrogen," *International Journal of Multiphase Flow*, vol. 130, p. 103360, 2020.

[2] M. Alamgir and J. Lienhard, "Correlation of pressure undershoot during hot-water depressurization," *Journal of Heat Transfer*, vol. 103, no. 1, pp. 52–55, 1981.

[3] J. Bartak, "A study of the rapid depressurization of hot water and the dynamics of vapour bubble generation in superheated water," *International Journal of Multiphase Flow*, vol. 16, no. 5, pp. 789–798, 1990.

[4] Y. Moshkovich, Y. Levy, and E. Sher, "Theoretical criteria for homogeneous flash boiling atomization," *Chemical Engineering Science*, vol. 206, pp. 471–475, 2019.

[5] J. Lienhard, M. Alamgir, and M. Trela, "Early response of hot water to sudden release from high pressure," *Journal of Heat Transfer*, vol. 100, no. 3, pp. 473–479, 1978.

[6] A. R. Edwards and T. P. O'Brien, "Studies of phenomena connected with the depressurization of water reactors.," vol. 9, pp. 125–135, 1970.

[7] A. M. Log, M. Hammer, H. Deng, A. Austegard, A. Hafner, and S. T. Munkejord, "Depressurization of co2 in a pipe: Effect of initial state on non-equilibrium two-phase flow," *International Journal of Multiphase Flow*, vol. 170, p. 104624, 2024.

[8] V. P. Carey, *Liquid-vapor phase-change phenomena: an introduction to the thermophysics of vaporization and condensation processes in heat transfer equipment.* CRC Press, 2020.

Eulerian Sectional Approach for Particle Erosion in Compressible Flows

Amir Loyevsky[1,2], Ido Immer[2] and Yuval Dagan[1]

[1] Faculty of Aerospace Engineering, Technion - Israel Institute of Technology, Haifa, 3200003, Israel

[2] Aeronautical Systems (39), RAFAEL, Haifa, 31021, Israel

Particle dispersion and erosion play an important role in various engineering applications, particularly in impinging jet flows. This study extends a recently developed Eulerian framework for characterizing these phenomena in such jets, where the framework can now handle bisectional systems. The Eulerian mass and momentum conservation equations are solved, assuming a one-way coupling between the flow and the particles. Erosion at the eroded surface is computed using an empirical erosion model. The investigated flow configuration allows estimating the initial erosion for bisectional systems by first evaluating the initial erosion in the corresponding monosectional systems. Then, the erosion induced by particles from different sections is linearly added using a simplified calculator with different weights for each section. This calculator generates the initial erosion immediately and spares the need to run numerous expensive simulations for different bisectional systems. Analysis of the monosectional systems showed that smaller Stokes number particles experienced smaller Stokes drag, following the streamlines of the carrier flow as they were radially diverted from the axisymmetric jet's core; medium Stokes number particles were diverted slightly from the axis of symmetry before impinging the wall, thus creating a W-shaped erosion profile, and the largest Stokes number particles experienced the largest Stokes drag, and therefore, induced a U-shaped erosion profile. Analyzing the normalized recession magnitude showed that our results stand in agreement with the guidelines used in the field for the design of materials in the converging section of rocket engines. Analysis of the bisectional systems showed that intermediate erosion profiles are obtained when using different ratios between medium and large Stokes number particles. This numerical model offers a computationally efficient framework for CFD-based transient erosion calculation.

Manuscript 31 submitted to 64th Israel Annual Conference on Aerospace Sciences (IACAS). Received December 30, 2024.

Trajectory Design and Control for Missed-Thrust Rendezvous

Meir Nemirovsky and Pini Gurfil
Department of Aerospace Engineering
Technion -- Israel Institute of Technology, Haifa 3200003, Israel

Missed thrust events (MTEs) pose significant challenges for rendezvous missions, jeopardizing spacecraft safety and mission success. Historical mission data suggests that servicing spacecraft equipped with low-thrust electric propulsion systems may experience several MTEs annually, with some events potentially cascading into multiple failures. Recognizing MTEs as a frequent occurrence during servicing rendezvous procedure, this presentation introduces a resilient methodology for trajectory design and control to mitigate MTE effects, ensuring safe and reliable operations.

The proposed methodology incorporates a Missed Thrust Sensitivity Grade (MTSG) as a key metric to identify resilient trajectories during mission planning. This metric informs the optimization of initial relative conditions, maneuver duration, trajectory profiles, and controller parameters. An iterative process ensures compliance with operational thrust and state constraints while minimizing the expected impact of MTE-induced deviations.

During maneuver execution, a closed-loop controller is employed to correct timing and state errors. Furthermore, an active terminal strategy based on rendezvous guidance laws replaces conventional passive coasting, enhancing terminal accuracy and improving resilience to MTEs.

The study examines various MTE profiles and suggests resilient trajectory designs for low-eccentricity targets in both LEO and GEO orbits. Simulations leveraging linear and nonlinear dynamic models confirm the framework's robustness, achieving a high success rate of approximately 97% for thrust losses up to 8% of total maneuver duration. Even in severe cases with thrust losses exceeding 37%, the methodology ensures mission success and precise maneuver execution within the desired time frame. This approach provides a practical, reliable solution for rendezvous missions facing unpredictable thrust loss scenarios.

An Optimal Low-Thrust Spacecraft Interception Guidance Law with Terminal Velocity Constraints

Yahli Drucker*, Vitaly Shaferman†
Technion, Israel Institute of Technology, Haifa, Israel, 3200003

Submission for the Israel Annual Conference on Aerospace Science (IACAS) 2025

Abstract

Satellites play a significant role in many important activities, such as communication, navigation, scientific research, and even intelligence gathering. As the number of satellites in space rapidly increases, the missions of protecting and disrupting them have become highly important. Hence, the problem of spacecraft interception with terminal velocity constraints using low thrust is considered in this work. It is assumed that both spacecraft are relatively close to each other and that the target spacecraft is in a circular orbit, so the problem can be formulated with the linear Clohessy-Wiltshire dynamics equations. The optimization problem is posed using a quadratic cost function containing a running cost on the control effort, and a terminal cost on the relative position and the difference between the relative velocity and a user-defined commanded velocity. Thus, a linear-quadratic optimization problem is formed, with soft constraints on the relative position and velocity. To simplify the problem, the zero-effort-miss state transformation is used. This approach enables the derivation of analytical, explicit, open-loop and closed-loop solutions using Pontryagin's Minimum Principle. Based on these optimal solutions, a second, static optimization problem is formulated. The objective function of the new optimization problem is the optimal cost of the main problem, and the optimization variables vector is the commanded velocity. Two different requirements are considered by applying two different constraints on the commanded velocity vector: in one case, a norm constraint is applied such that the command terminal speed becomes a user-defined parameter and the direction of the velocity is optimized, while in the other case, the terminal velocity direction is predefined and the terminal speed is optimized. Analytic solutions for the command velocity are obtained for the case where the direction is constrained and for the case where the speed

*Graduate Student, Faculty of Aerospace Engineering, Technion, yahlidrucker@campus.technion.ac.il.
†Associate Professor, Faculty of Aerospace Engineering, Technion, vitalysh@technion.ac.il.

Manuscript 24 submitted to 64th Israel Annual Conference on Aerospace Sciences (IACAS). Received December 23, 2024.

is constrained and assumed to be high. Furthermore, the command velocity can be found for any constrained speed by solving for the roots of a sixth-degree polynomial. By substituting the resultant command velocity into the optimal solution of the main optimization problem, the optimal guidance law for each case is obtained. The performance of the derived guidance law is evaluated in simulations in four different cases: without commanded velocity constraints, with a constrained commanded speed (both high and low speeds, relative to the speed obtained in the unconstrained case), and with a constrained velocity direction. The solutions in all four cases are compared to the GPOPS-II direct optimization solutions. In addition, the optimality of the constrained commanded velocity is evaluated graphically. It is shown that the solutions of the proposed guidance law satisfy the constraints with minimal cost and agree with the optimal numeric solutions.

Deep Space Navigation using Satellite Based Radio Interferometry

Moshe Golani, Prof. Yoram Rozen, Dr. Hector Rotstein

Abstract

Orbit Determination (OD) in deep space (beyond the far side of the Moon) presents significant challenges due to several factors. First, Global Navigation Satellite Systems (GNSS) signals are unavailable at such distances, so that alternative navigation methods are required. Second, the spacecraft's beacon transmitter signals experience significant propagation losses over long distances, arriving at Earth-based receivers with considerable noise and weakness. Furthermore, geometrical triangulation is hindered by Earth's small size relative to the target's distance.

NASA and ESA primarily rely on the Deep Space Network (DSN) for deep space exploration. The DSN is a terrestrial system with large antenna arrays of up to 70 meters in diameter distributed globally, has demonstrated nanoradian-level interferometric accuracy. However, this system faces challenges, including short baselines for deep space, atmospheric scintillation requiring extensive calibration process, and limited Line-of-Sight (LOS) due to Earth's rotation.

This research builds on the strengths of Very Long Baseline Interferometry (VLBI) and proposes a system model utilizing Geosynchronous Equatorial Orbit (GEO) satellites as differential phase measurement stations. These GEO satellites serve as space-based interferometric OD platforms, with phase data logged, time-stamped, and transmitted to ground stations for processing. The proposed method offers extended base-lines, improving accuracy for a given sensor size, while eliminating atmospheric errors such as ionospheric and tropospheric distortions and enhancing system availability. Simulation results of this concept, named as Radiometric Interferometry for Navigation using GEO Satellites (RINGS), show improved system availability in approximately factor 2, improved noise scheme for solar missions, and promising potential to perform radio interferometry with relatively small size antenna mounted at a GEO satellite. Challenges of this concept include clock synchronization and accurate GEO satellite location awareness to obtain the desired sensor accuracy for deep space navigation.

Manuscript 76 submitted to 64th Israel Annual Conference on Aerospace Sciences (IACAS). Received January 11, 2025.

A Space Propulsion System without Mass Ejection

Azriel Lorber*

Abstract

Flight performance of long-range spacecraft is limited by the need to lift large amounts of fuel into orbit; thus, a propulsion system which will not depend on mass-ejection will be a most welcome development. Various past proposals which attempted to achieve this were rejected because it was thought that they contradicted basic laws of physics.

An experimental mechanical device which operates without mass-ejection, (or interaction with the environment), and which can serve as a spacecraft propulsion system, was constructed and **successfully** tested. It is based on manipulating centrifugal forces, internally generated by rotating masses, and is not size or scale dependent.

Full explanation, with photos and movies of its construction and operation, is presented. When fully developed, this technology will enable constant long-duration acceleration of spacecraft (decreasing flight times), permit novel physics research, or enable hovering with or without the presence of atmosphere.

Key Words

Ejection-less propulsion, Reactionless space drive, Space propulsion systems, Space travel,
Long-range spacecraft

I. Introduction

Currently, spacecraft designed for long missions are lifted into orbit, accelerated to the desired speed (usually by mass ejecting chemical rockets which have Isp of up to 450 seconds, depending on type) and then "coast" to their destination. Since the fuel for such journeys must be lifted too into orbit, long range spacecraft are limited in their maximum velocity, and this bears directly on transit times with all attendant problems of long space voyages. If such a spacecraft is to orbit another celestial body, more fuel is required for deceleration, when it approaches its destination.

Nuclear-powered rockets which work by heating and ejecting stored Hydrogen were investigated too. Although more efficient than chemical rockets, (Isp of 1400 to 2000 seconds) they too eject consumables.

Various electric thrusters, which eject high-velocity particles, are currently studied. Compared to chemical and nuclear-powered rockets their specific impulse is high, 2500 to 4000 seconds, but the weight of the ejected mass is small and consequently the forces they produce are minute, on the order of fractions of Newtons. This suffices for station-keeping of satellites and similar applications but not for long distance travel.

Obviously, a propulsion system, which will not depend on mass discharge, will be a most desirable development for future space applications.

II. Previous work

In the past, many inventors attempted to design devices which will create a lifting force, essentially a form of "anti-gravity", without the ejection of mass, and without utilizing propellers. These devices were usually based on asymmetric pulses of oscillating or rotating masses. With the advent of spaceflight, there were many attempts

* Independent researcher, Associate Fellow, AIAA; azriell@zahav.net.il

to adapt such concepts to space flight. Just writing "Reactionless Space Drive" in google brings up many such.

There are also hundreds of patents, patent applications and other proposals for such devices, from all over the world.[1]. Some of them are conceptually similar to the current work. Perusal of these designs shows that all suffer from two deficiencies: They are based on overly complicated mechanisms or on arrays of gears, pulleys and cables, both of which will make them difficult to construct and unreliable to operate. Secondly, none of these publications present or depict actual working models - only drawings and explanations of the concepts.[2]. What's more, even in publications which analyze most advanced (albeit theoretical) propulsion concepts, there is little discussion, always disparaging, of such technologies. [3].

A few papers on various aspects of this subject were presented at scientific meetings and journals. [4, 5, 6]. In view of these technologies' importance and potential, especially for space travel and research, it is surprising that they were not further developed, at least to a prototype or at least proof-of-concept level.

Part of the reason for this might be the resistance of some of the scientific community, which claims that it is impossible to achieve a propulsive force without mass ejection or interaction with the environment. It is stated that tests which claim to achieve this were carelessly conducted. [7]. Others insist that such ideas violate basic laws of physics. [8, 9, 10]. Also, some non-technical objections were raised regarding work in this field. [9]. It is possible that this attitude dissuaded researchers from further work on such concepts.

III. Explanation of the System's Design and Operation

Despite such objections, a working model of such a system was designed, constructed and successfully tested. This paper describes the device's construction, operation and test results.

Nomenclature: (See also Fig. 1.)

F – Centrifugal force exerted by masses m on the wall A-B-C-D-A

R_1 – Longer radius, defined between points A and B.

R_2 – Shorter radius, defined between points C and D.

F_1 – Centrifugal force along wall from A to B.

F_{T1} – "North"-pointing integral of F_1 along wall from A to B.

F_2 – Centrifugal force along wall from C to D.

F_{T2} -- "South"-pointing integral of F_2 along wall from C to D.

F_{net} – F_{T1} - F_{T2}; "North" acting net force (in the two-armed configuration).

m – Lead masses mounted on arms. (See Fig. 1 and Images1, 3,4 and 5.)

n – Angular velocity of the masses in rounds per second

V - Linear velocity of the masses along the wall.

R^2 - (In Fig. 5) Coefficient of determination.

(The length of the arms carrying m varies with their angular positions.)

The basic idea was that the magnitude of a centrifugal force could be manipulated along its trajectory, to generate a unidirectional, controllable force. See Fig. 1.

The system consists of lead masses (m) mounted at the ends of telescoping (but free to slide in and out) arms. The momentary length of these arms, and hence the masses' orbits and **linear** velocity along the wall, are defined by the shape of the limiting wall and their angular position. See Image 1. The concept of the limiting wall was described in [2] although without delving into design details.

Basic equations for a twin-armed model (see Fig. 1).

Centrifugal force equation: $F = mv^2/R$ (1)

and $V = 2n\Pi R$ (2)

resulting in $F = m4^2n^2R$ (3)

and thus $F_1 = m4\Pi^2n^2R_1$ (4)

$F_2 = m4\Pi^2n^2R_2$ (5)

F_{T1} and F_{T2} are integrals of F_1 and F_2 along respective sections of the wall; see the Nomenclature and Fig. 1.

Fnet (acting "North") = $F_{T1} - F_{T2}$ (6)

(Forces between BC and DA are essentially cancelled by the symmetry of the system.)

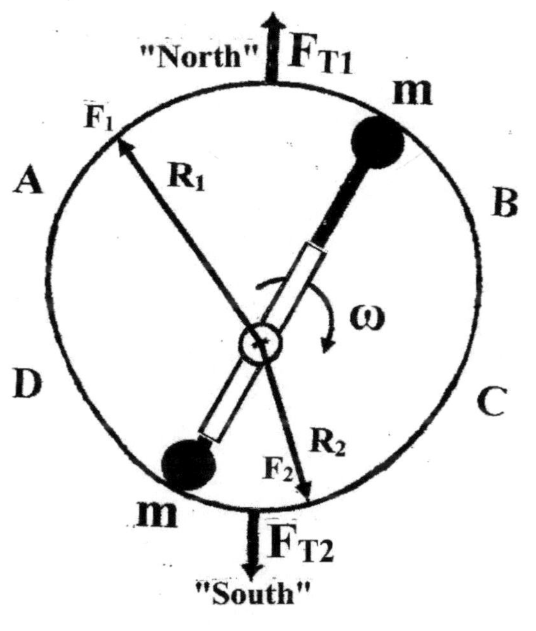

Figure 1: A Schematic of a Twin-Armed System

IV. The Test Model

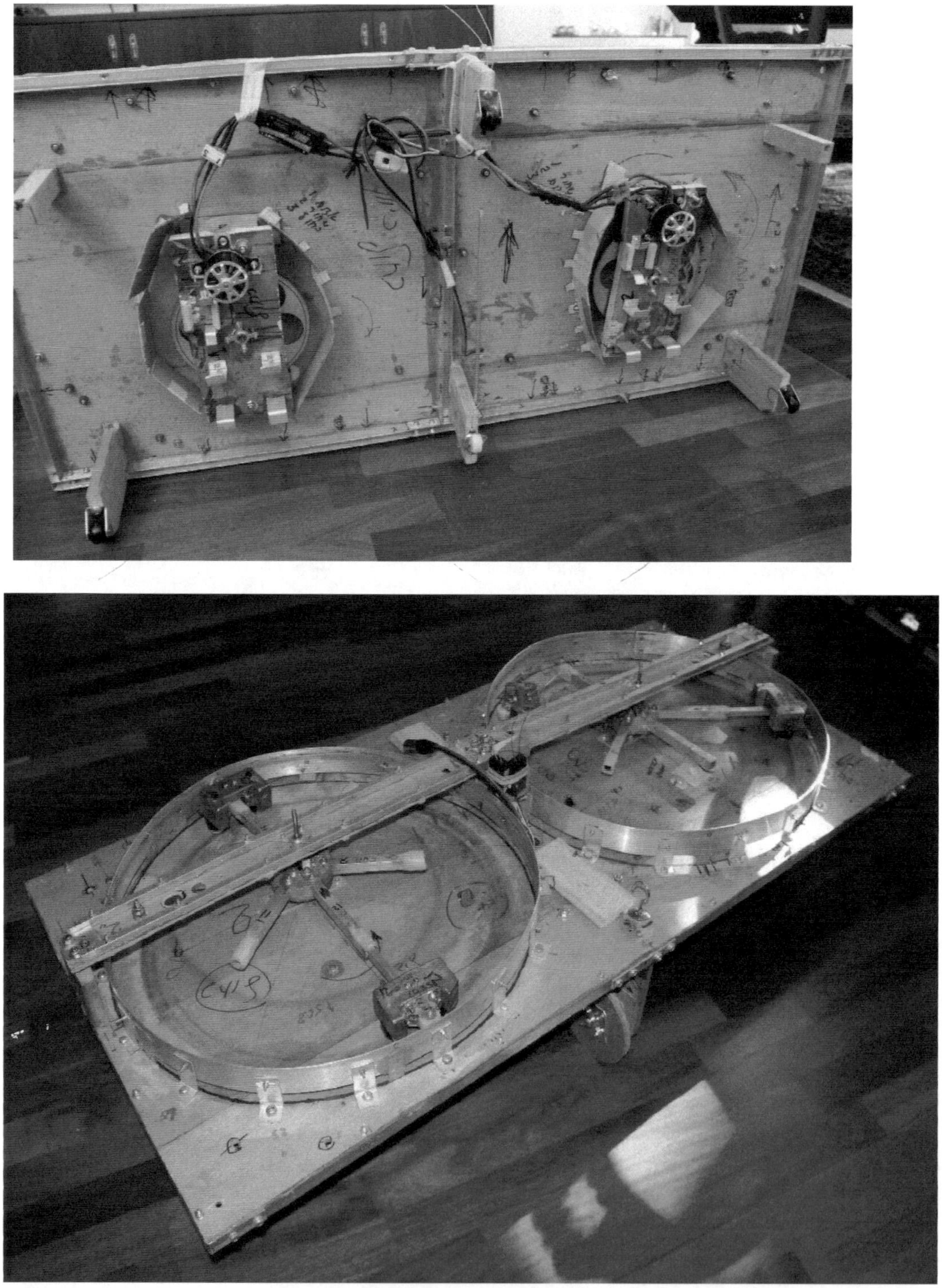

Image 1: **Layout of a Twin Armed Test Model:** to eliminate rotational moments of the motors the system consists of two similar devices rotating in opposite directions. Initially it

was planned to operate with lead masses on all six arms. Turned out that the motors couldn't handle more than two of the masses. The model shown is the latest of several design improvements.

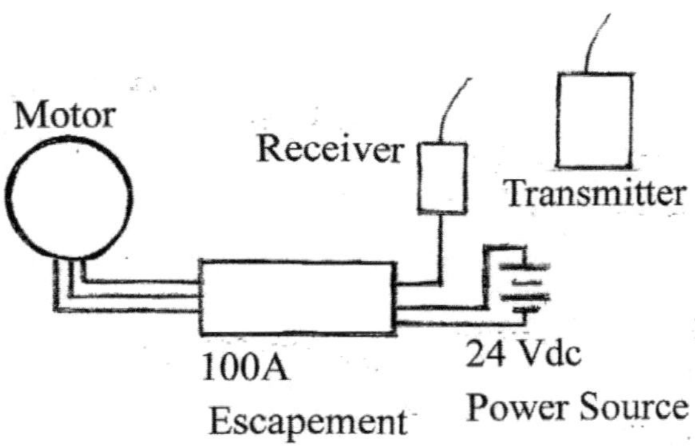

Fig. 2: A Schematic of the Model's Electrical System for each side.

Image 2: An Underside View of the Test Model. (The four unpowered wheels in this photo were later replaced by larger ones, unpowered too, as in image 1.) The large gears serve to reduce the motors' angular velocity. See also **Image 6.**

System's Data and Test Results

(The test article was built in a home-workshop using only basic tools.)

Motors -------------------------------------	Brushless, TomCat G90 5625-KV330
Gear ratio, motors to rotating arms ----	11.
Control system --------------------------	2.4 GHz Futaba transmitters and receivers with RC models 100A escapements; see Fig.2.
Digital Force Gauge ---------------------	ALIYIQI Instrument Co. Ltd. Model AMF-100
Photo/Contact Tachometer --------------	Model DT-2268, Lutron (USA).
Platform's weight without batteries ----	15.040 Kgs (Single lead mass on each side.)
Power for motors ------------------------	24V_{DC} (two 12V batteries)
Weight of battery pack ------------------	5.285 Kgs
Total weight of platform ----------------	20.325 Kgs (Single lead mass on each side.)
R_1 of the wall between A and B --------	0.315 meter in the current model.
R_2 of the wall between C and D --------	0.225 meter in the current model.
R_1m – rotation radius of mass -----------	0.26 meter*
R_2m - rotation radius of mass -----------	0.17 meter*
Individual lead masses' (m) weights ---	1.1± 0.1Kg each.**
ω Angular velocity in current tests ------	200 – 750 RPM
Platform's length (in direction of movement) -----------	0.6 meter

Platform's width -------------------------- 1.25 meter

*These values are for the rotation radii of the centers of the masses along sections A-B and C-D, respectively.

** These numbers are for the latest configuration, performance of which is given in Table 1 and Fig. 5.

V. Test results

The movement tests (recorded in the supplementary movies – listed at the end of the paper) were performed with either one operating arm (with a mass at its end) or with two. With one arm the force F_1 is larger than F_2 (see **Fig. 3a**) but the result is a more pronounced periodicity of the system's movement. (Movie-clips 1 & 2.) With two arms the "North" pointing total force (F_1-F_2) is smaller, twice as frequent (see **Fig. 3b**) but more stable. (Movie-clip 3).

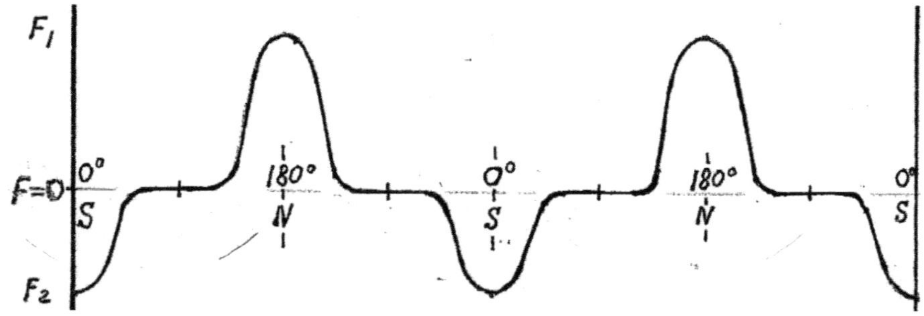

Fig. 3a: Forces Along the Circumference of the Device for the One-Armed Configuration. Starting point for this drawing is at the "South" point. F1 points "North" and F2 points "South".

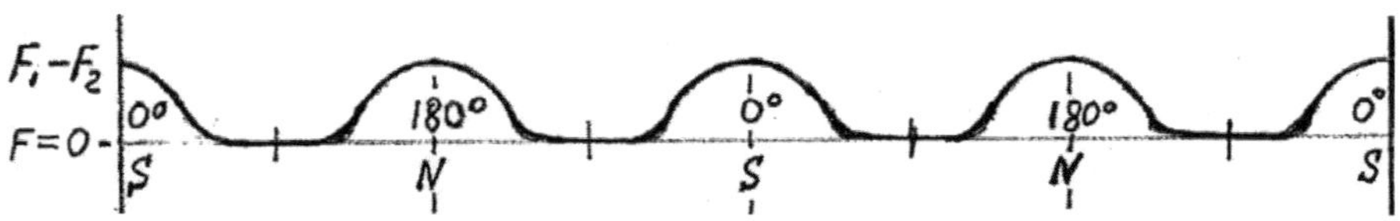

Fig. 3b: Forces Along the Circumference of the Device for the Twin-Armed Configuration. Starting point is at the "South". F1-F2 points "North".

These forces, F_1 and F_2, were measured (see Fig. 4) and calculated, and the results are presented, separately for F_1 and F_2, in Table 1 and Fig. 5.

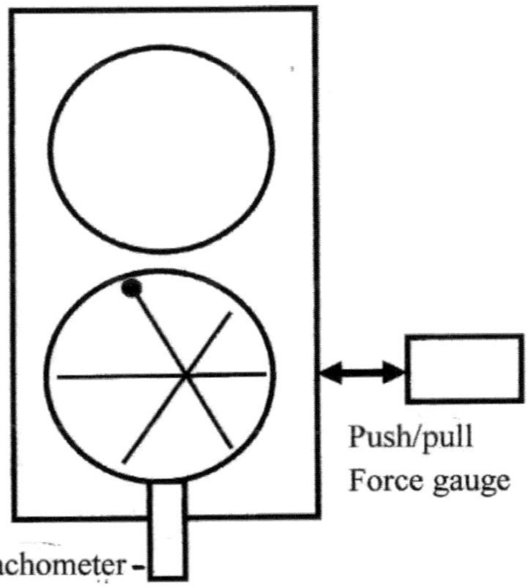

Fig. 4: F_1 and F_2 measurements set-up.

(F_1 was measured pulling to the left, as in this figure; For F_2 measurements the platform was turned around and F_2 too was measured pulling to the left; This because pushing measurements posed some difficulties and because this procedure avoided calibration differences (in the force gauge) between pull and push.)

Table 1: F_1 and F_2 calculated and measured test results as function of RPM

RPM	F_1 calculated Kgs	F_1 measured Kgs	RPM	F_2 calculated Kgs	F_2 measured Kgs
220	0.421	0.4	250	0.355	0.26
320	0.890	1.1	340	0.657	0.65
380	1.250	1.5	440	1.10	1.3
420	1.530	1.9	610	2.11	2.0
490	2.090	2.3	620	2.18	2.4
520	2.350	2.7			
600	3.130	3.45			
620	3.350	3.65			
650	3.750	---			
750	4.900	4.88			

See also Figs. 3a and 5.

The calculated results were obtained using equation 3 from section III. The procedure used was
as follows:

1. The measured RPM was divided by 360. (The Tachometer counted all 6 arms.) This gave RPS.
2. The resultant RPS was squared and multiplied by $m4\Pi^2R$. (m was derived from the masses' average weight of 1.1 Kg. See System's Data.) This multiplier was 1.13 for F_1 and 0.737 for F_2. The results of these gave the calculated results.
3. The measured F_1 and F_2 were read manually from the force gauge; see Fig. 4.

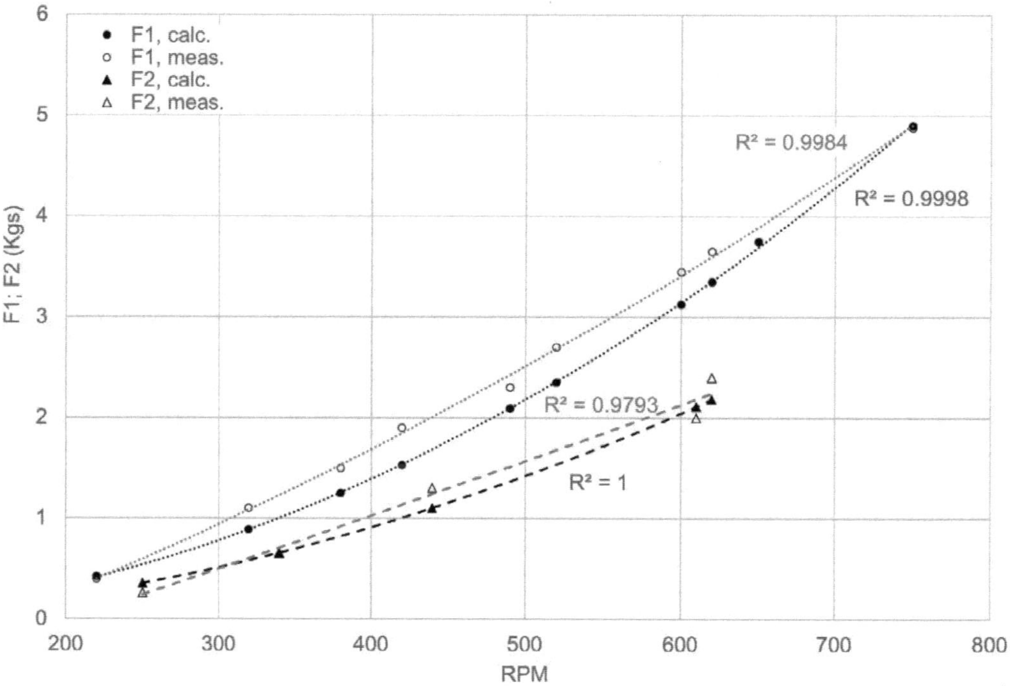

Fig. 5: Measured and Calculated F_1 and F_2 Peak Forces. (See also Fig. 4.)

VI. Discussion

1. This work shows that an object can be accelerated without mass ejection (as in rockets) and with no interaction with the environment. The forces to achieve this were generated by internal electrical motors. A physical model was constructed and tested for this purpose.
2. The movement of this model – the platform – which can be seen in the supplementary movies, was achieved by two similar modes of operation: one, a unidirectional force, F_{T1} (which is bigger and thus overcomes F_{T2} in the one-armed configuration) or Fnet (F_{T1}-F_{T2}) in the two-armed configuration. These forces act in the desired "North" direction and are generated by the rotating masses. The difference in the magnitude and timing of these forces (on both sides of the model) generated by one or two operating arms is shown in Figs. 3a and 3b. This difference and timing affect the dynamics of the platform's motion, as demonstrated in the supplementary movies.
3. A common argument against the presented concept is that it violates the Law of Conservation of Momentum, though without explaining in what way it does so. To start with, this law is not applicable in the current case. See [11].
   Furthermore, in this case, the angular momentum of the rotating masses is converted into the linear momentum of the platform. The platform starts with zero momentum (zero velocity) but the centrifugal forces (generated by the internal electric power) cause it to move and gather velocity (and thus momentum) without ejecting mass.
4. Another argument against the validity of such concepts is that the platform's movement is impossible without ground friction. [7].

Newton's Second Law states that if a force is acting on a body, the body will accelerate. Since it was already demonstrated that such forces are generated and do exist, the platform obeys Newton's Second Law without the need to involve ground friction.

5. Furthermore, for the argument about the ground friction to be true, this friction must act as a one-way ratchet, blocking backward movement, and allowing it in one direction only. [7]. Clearly, this is not so. In the one-arm mode, the platform initially reacts to the forces generated in both directions, but because the "North" acting force is bigger than the one to the "South" (Fig. 3a) the platform's direction of movement is quickly stabilized, moving steadily "North". (Supplementary movies 1 & 2). In the twin-armed configuration, forces F_{net} act only in the "north" direction (Fig. 3b) and the platform moves north, without seesawing. (Supplementary movie 3).
   Ground friction will come into effect only in powered wheels, and without it (like in a car stuck in mud) there is no movement. Since the current wheels are unpowered, the ground can't act as a reaction mass, as suggested in [10] and (according to Newton's Second Law) this will be true also in interplanetary space.
6. According to Newton's Third Law, for every action there is a reaction. It was claimed that since such a reaction is absent in the current case, the proposed concept is invalid.
   It should be remembered that while rockets do move as a reaction to mass-ejection (Newton's Third Law) there is no law that states that movement <u>requires</u> mass ejection.
   Also, when the lead masses push against the wall and generate a force (see Table 1, Fig 5 and Images 4 and 5) the wall pushes back against them, or to be exact, against the little wheels at their ends. (see image 3 and 5). This is "Third Law" in Action. But the wall's pushback exerts this force only on the lead masses, and because they are free-standing (they can slide freely in and out of the arms) this reaction force (the pushback) does not affect the rest of the system
7. Due to the strong effect of the masses' angular velocity on the system's performance (see equation 3 in section III) increasing this velocity to the likes of internal combustion engines, will produce forces of thousands of Newtons.

VII. Conclusions

A prototype mechanical device which produces a unidirectional controllable force without mass ejection, and with no interaction with its environment, was designed, constructed, and **successfully tested.**

The current work is a feasibility study, intended to serve only as a proof of concept, but it proves conclusively that a Reaction-Less-Space-Drive is possible.

More development work is needed to fully realize this technology's potential, future applications of which include:

1. Constant acceleration of space vehicles for extended times, to reach higher velocities. This will shorten flight durations to far destinations - "To Mars and back in twenty days" is a realistic goal.
2. Constant acceleration for very long time will enable research in novel fields of physics.
3. Directed upward this technology will enable hovering with or without atmosphere. This can be used to design more survivable, rotor-free helicopters.

4. Also, directed upward, this technology may replace, or supplement, launching rockets.

VIII. Future Work

Fig. 1 and 2, and Images 1 – 5 give enough details for a technically competent person to be able to copy the design and perform the tests.

The first step in further developing this technology will be to professionally design and construct a model of the system (or a set of such) capable of withstanding constant rotational speeds of several thousand RPM and forces of hundreds, possibly thousands, of Newtons. This, in order to test and optimize various design parameters, such as overall size, the number of operating arms, (definitely should be increased, with suitable motors), best wall's geometry, ratio of R_1/R_2 and optimum weights of masses, all of these to maximize forces (F) generated, against power requirements.

The next step can be the testing of such a device with forces acting horizontally (without wheels) or vertically - upward. For all such tests a carefully balanced system is a must.

Image 3. A Detail of Image 1.

Image 4. Another Detail of Image 1.

Image 5: Another Detail of Image 1 (The photos were taken from successive designs of the platform, hence the different shapes of the lead masses.)

Image 6. A Detail of Image 2.

IX. End Notes

[1] It will be impossible to list here all the patents in this field. A partial list of relevant ones can be accessed at "US Propulsion Patents 1974 – 1989", which also quotes previous works, and in addition lists several foreign patents in this field. A more recent one is [2] below and some are listed at the end of [7].

[2] See US Patent 20150314868A1, filed in 2013, A Device for the Generation of Lift, by Angelo Raffaele Alterio. This patent application was singled out because it proposed an **idea** similar to the one in the current paper, but without any design details, which makes it impossible to analyze or duplicate. (This patent however was abandoned.)

[3] Millis, Marc, G. and Davis, Eric, W., (eds.) "Frontiers of Propulsion Science", AIAA Progress in Astronautics and Aeronautics", Volume 227, 2009.

[4] Yu, H., Yang, T., Liu, Y. and Wane, S., A Further Study of Control for a Pendulum-Driven Cart, International Journal of Advanced Mechatronic Systems, (2008) Volume 1, Issue No 1, pp. 42-52.

[5] Loureiro, J., Alcaso, A., Pitrama, R., Design and Development of Propulsion Systems Without Traction Wheels, Propellers, Jets, or Rockets, Proceedings of the 7th Conference on Mechanics and Materials in Design, Albufeira/Portugal, 11-15/6/2017, Paper Reference 6853.

[6] Burns, David, M., Helical Engine, https://doi.org/10.2514/6.2019-4395.

[7] Millis, Marc, G., Nonviable Mechanical "Antigravity" Devices, in Millis, Marc G. and Davis, Eric W. (eds.) Frontiers of Propulsion Science, in AIAA Progress in Astronautics and Aeronautics, Vol. 227, pp. 249 – 261, (2009). Unfortunately, in the second paragraph on page 251, the author made an unwarranted assumption concerning ground friction (later proven wrong) which renders his analysis somewhat dubious.

[8] Private communications.

[9] Millis, Marc G. and Thomas, Nicholas E., Responding to Mechanical Antigravity, NASA/TM – 2006-214390 (AIAA – 2006 – 4913), December 2006, pp.1- 4.

[10] Millis, Marc, G., Prerequisite for Space Drive Science, in Millis, Marc, G. and Davis, Eric, W., (eds.) in "Frontiers of Propulsion Science", in AIAA Progress in Astronautics and Aeronautics", Volume 227, p. 128, (2009).

[11] The Law of Conservation of Momentum states that "in an isolated system, the total momentum of two or more bodies acting upon each other remains constant unless an external force is applied". Another version of the law analyzes collisions between such bodies. The current system is a closed one, consisting of a single body and no collisions.

List of Figures

List of Images

List of Tables

List of Supplementary Movie-Clips of Several of the Tests

1: Test of a One-Armed System.
https://drive.google.com/file/d/1I5PXuUfjSh-N9czkGhx-_xE8OF_9hQ0k/view?usp=drive_link
2: Another Test of a One-Armed System.
https://drive.google.com/file/d/1ZBcZq_q0HKvZYPUKperhEiBt_vZgw-K5/view?usp=drive_link
3: Test of a Twin-Armed System.
https://drive.google.com/file/d/1Y63bROkXhPSpQ9sbM4ZF4pJR_qB-AxF2/view?usp=share_link

Thanks

The author wishes to thank his children, Dana and Itai, and his wife Nurit, for their advice and practical help in various aspects of this system's development.

An Optimal Soft Landing Guidance Law with an Approach Angle Path Constraint

Revital Frenkel * and Vitaly Shaferman †
Technion - Israel Institute of Technology, Haifa 3200003, Israel

Abstract

The challenge of achieving a soft landing has been a significant research focus for many years. The term soft landing refers to the controlled and gradual descent of a spacecraft to a designated landing site, minimizing impact forces upon contact with the surface (i.e., zero velocity at the designated position). Optimal control-based solutions are very appealing as they can produce very low miss distances and velocity misses, minimal propellant consumption, minimal time, or any other desired criteria. The known optimal control-based solutions to the soft landing problem can address many practical scenarios. However, some solutions may produce undesired and dangerous outcomes, such as ground collisions and extremely low approach angles. Several studies have addressed these issues, but most of the suggested solutions have substantial numerical components or are totally numeric. Analytical solutions to this problem are preferred because they are usually faster and easier to implement in real spacecraft with limited computational power. Furthermore, they allow for the prediction of performance and the exploration of different aspects of the solution, facilitating a better understanding of the expected results.

In this study, a new optimal-control-based soft landing guidance law with a relaxed inequality path constraint (i.e., transition from an unconstrained arc to a constrained arc without leaving the constraint) is analytically derived. The proposed guidance law prevents ground collision and allows approach angle control by constraining the optimal trajectory to remain above or on a minimal approach angle line, originating from the landing point, in addition to the miss distance and miss velocity soft constraint at the end of the scenario and the minimization of the control effort. A 2D point mass dynamics with a constant gravitational field is employed, along with a linear dynamic model and a quadratic cost function. Analytical open-loop and closed-loop solutions are derived using Pontryagin's Minimum Principle and the transition optimality conditions between the unconstrained and constrained arcs. This guidance law is then used to formulate a simple algorithm for determining an optimal trajectory confined within a triangular region defined by two lines originating from the landing point. The proposed guidance law and algorithm were evaluated in simulation for various initial conditions, and demonstrated very good performance and satisfaction of the path constraints.

*Graduate student, Faculty of Aerospace Engineering. E-mail: revitalfr@campus.technion.ac.il
†Associate Professor, Faculty of Aerospace Engineering. E-mail: vitalysh@technion.ac.il.

Manuscript 73 submitted to 64th Israel Annual Conference on Aerospace Sciences (IACAS). Received January 9, 2025.

Exploring Low-Cost Trajectories to the Sun-Earth Lagrange Point

Geffen Aharoni[1], Tamar Alperin[1], Naama Gilat[1],
Iris Kanter[1], Almog Yanku[1],
Alex Frid[1,2]

[1]Faculty of Aerospace Engineering
[2]Asher Space Research Institute (ASRI),
Technion - Israel Institute of Technology, Haifa, Israel
geffen-aha@campus.technion.ac.il, tamar.al@campus.technion.ac.il,
naamagilat@campus.technion.ac.il, iriskanter@campus.technion.ac.il,
almog.yanku@campus.technion.ac.il, afrid@technion.ac.il

Abstract

Earth's warming is an existential challenge necessitating innovative solutions to mitigate its impacts. One such solution is the deployment of a swarm of solar sails at the Sun-Earth Lagrange Point (SEL1) to regulate solar radiation reaching Earth. However, reaching SEL1, located approximately 1.5M km from Earth, presents significant financial and technical challenges due to the high energy demands of traditional propulsion methods. This work explores novel, cost-effective strategies to overcome these barriers for a cost effective technology demonstrator and as well as for the full scale solution, focusing on leveraging solar sails, "ride share" optimization and direct trajectory optimization.

Our approach includes developing a simulation platform for evaluating and optimizing mission cases for mission trajectories:

Weak Boundary Strategies: The Weak Boundary Condition (WBC) leverages the SE L1 point and the Weak Stability Boundary (WSB) to enable efficient trajectories. By utilizing gravitational interactions in a multi-body system, the WBC reduces fuel requirements by enabling spacecraft to transition into low-energy regions. This approach involves guiding the spacecraft into a highly elliptical orbit and using the natural dynamics near the WSB to minimize propulsion needs. Furthermore, this option can be used as a "ride share" opportunity with other missions, such as lunar missions, to optimize fuel usage and reduce launch costs. Once aligned, the spacecraft can transition towards the Moon's vicinity or further adjust to achieve the desired trajectory towards L1.

Manuscript 91 submitted to 64th Israel Annual Conference on Aerospace Sciences (IACAS). Received January 15, 2025.

Solar Sail Maneuvers: Initiating from a circular orbit around Earth, increasing the apogee opposite to the Sun, and strategically aligning with Earth's orbital path to utilize solar radiation pressure effectively. By waiting for half an Earth revolution, the satellite's apogee, initially pointing towards L2, will realign to point towards L1, achieving the necessary orbital parameters. **Inverse Simulation Techniques:** Understanding propulsion requirements through simulations of the restricted three-body problem (3BP). By simulating trajectories backward in time, we identify feasible paths from Earth's sphere of influence to L1, constrained by energy conservation principles. This inverse approach highlights optimal launch velocities and minimizes unnecessary parameter exploration.

Preliminary results demonstrate the feasibility of achieving L1 insertion using minimal propulsion, reducing mission costs and complexity. Our simulations provide insights into the velocity ranges and trajectory profiles required for these missions, forming the foundation for future propulsion system designs.

Manuscript 91 submitted to 64th Israel Annual Conference on Aerospace Sciences (IACAS). Received January 15, 2025.

Effect of magnetic field on P- and S-waves in magneto-active polymers

*Ankush Yadav [1] and Pavel Galich [1,] **

[1] *Faculty of Aerospace Engineering, Technion - Israel Institute of Technology, Haifa, Israel*

Magneto-active polymers (MAPs), a category of smart materials, exhibit tunable mechanical properties under the influence of external magnetic fields. This research investigates how magnetic fields affect the propagation of P- and S-waves in isotropic MAPs. An enriched magnetoelastic energy function was employed to model the behavior of MAPs, resulting in explicit expressions for P- and S-wave speeds as functions of deformation, magnetic field strength, and enriched material constants. For an undeformed body, the speed of the in-plane S-wave (polarized in the direction of the magnetic field) depends on the applied magnetic field, while the speeds of the out-of-plane S-wave (polarized perpendicular to the magnetic field) and P-waves remain unchanged. To experimentally validate these theoretical findings, commercially available MAP filament (IronPLA) was used, and the desired P- and S-wave samples were prepared via fused filament fabrication (FFF) using a commercial 3D-printer. The pulse-echo method was employed to measure P- and S-wave speeds according to ASTM E494-20. P- and S-wave ultrasonic transducers (Olympus V101 and Olympus V151-RB) were used with a pulser-receiver device (Optel 2.0). Wave arrival times were calculated using a threshold-based method applied to consecutive echo signals. Controlled magnetic fields were applied using a GMW 3474 dipole electromagnet to study their effects on wave speeds. The findings demonstrated a strong dependence of the in-plane S-wave speed on the magnetic field, attributed to changes in the elastic modulus in the presence of the magnetic field. The in-plane S-wave showed greater sensitivity to the magnetic field compared to the P-wave and out-of-plane S-wave, emphasizing anisotropic behavior of elastic waves in the isotropic MAPs.

Keywords: Magneto-active polymers, P-wave, S-wave, magnetic field, IronPLA, elastic waves

Correspondence: galich.pi@technion.ac.il

Manuscript 50 submitted to 64th Israel Annual Conference on
Aerospace Sciences (IACAS). Received January 7, 2025.

An investigation into semi-stabilized unsymmetrical thin-walled structure

S.E. Katzeff[a]

[a] *Israel Aerospace Industries, Lod, Israel*

Abstract

In semi-monocoque aircraft primary structure, thin-walled metallic sections are fastened or bonded to external skin to carry load in both hoop and longitudinal directions. Although symmetrical sections are preferred, practical design constraints sometimes necessitate unsymmetrical sections. It is generally assumed that these geometrically unsymmetrical sections are structurally equivalent. This assumption stems from the concept that the member is considered stabilized; meaning that it is prevented from deflecting in the transverse direction through its attachment to the skin. Such simplification allows engineers to analyze the section as if it were symmetrical, using standard bending formulas to calculate stress, primarily along the global horizontal axis. The 'abstract' product of inertia I_{xy} is typically ignored, allowing the structure to be considered fully stabilized. However, this assumption is non-conservative. Even a stabilized unsymmetrical section requires an additional stabilizing transverse moment at its attachments to stabilizing members, such as aircraft skin. The non-conservatism becomes more pronounced in structures stabilized at only one outstanding flange—a condition termed semi-stabilized. In such configurations, fewer fasteners or reduced bonding are needed to apply the stabilizing transverse moment, potentially compromising structural integrity. The proposed investigation examines stringer reinforcement as an example where such non-conservatism can significantly impact both the static strength and fatigue life of primary structure. The research presents a straightforward methodology for reinforcing existing unsymmetrical, semi-stabilized structures to optimize their structural performance. Finite Element Analysis (FEA) validates the proposed method's assumptions and reveals flange efficiency phenomena in semi-stabilized sections—a characteristic typically associated only with curved beam structures. The study further proposes a hypothesis explaining this phenomenon and develops a semi-empirical equation to predict the flange efficiency factor for unsymmetrical, semi-stabilized structures under bending.

Keywords: Unsymmetrical thin-walled structure, Stabilized beam, Flange efficiency, Stringer reinforcement

1. Introduction

A typical example of aircraft primary structure—comprising skin, stringers, and frames—is illustrated in Figure 1. Where 'every kilogram counts', optimized load paths are designed that exploit any cross-section to its maximum efficiency. Aircraft primary structures that fail to effectively carry their intended load are considered 'parasitic' and are systematically eliminated during final design iterations or subsequent weight-optimization exercises. The most efficient geometric cross section is usually symmetric about one, or both, axes of symmetry. However, practical design considerations often necessitate departures from this engineering principle, most notably in longitudinal stringer configurations. Metallic stringer profiles vary widely, as shown in Figure 5, with manufacturers displaying distinct preferences: Boeing favors the symmetrical Hat-type profile, while Airbus generally employs the unsymmetrical Z-type stringer profile—the latter being the focus of the proposed investigation. The consequences of stringer profile selection are not always fully comprehended at the design's outset. Nevertheless, potential negative outcomes can be mitigated across various stages of the aircraft's lifecycle. Structural efficiency initially compromised during the design phase can be subsequently recovered through refined design or strategic repair approaches. Consequently, structural optimization and weight-saving considerations can be systematically applied throughout the aircraft's entire operational lifespan.

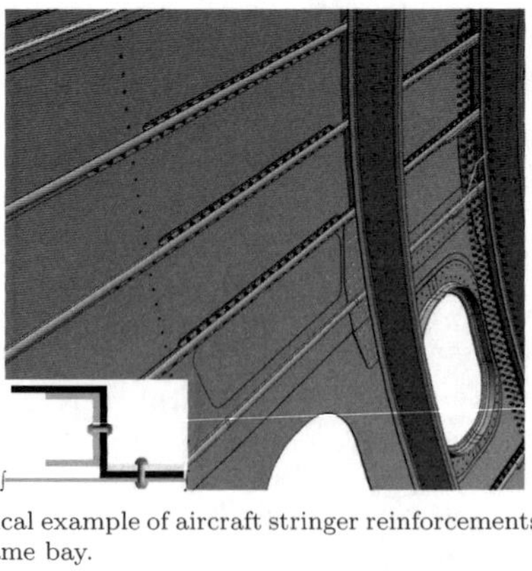

Fig. 1: Typical example of aircraft stringer reinforcements, extending over one frame bay.

*Data will be made available on request.

Email address: skatzeff@iai.co.il (S.E. Katzeff)

Preprint

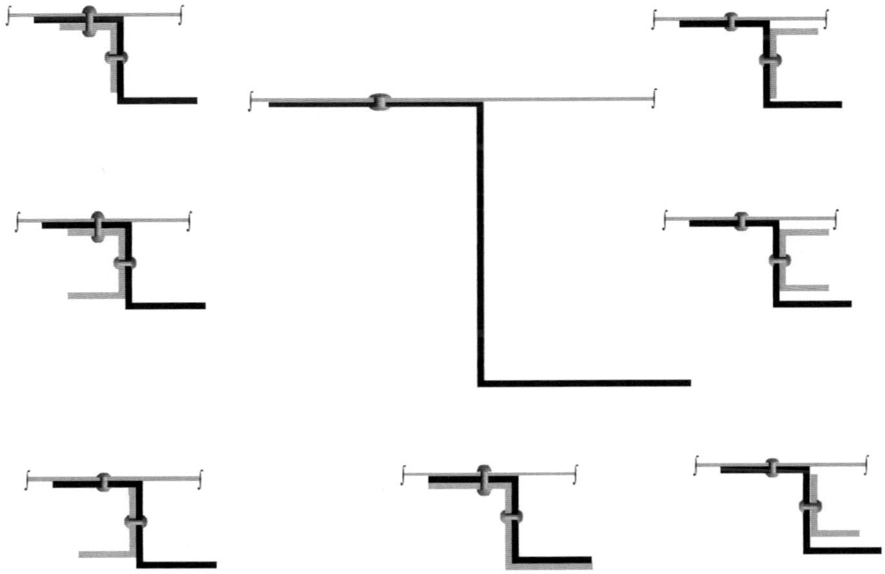

Fig. 2: Typical examples of stringer reinforcement.

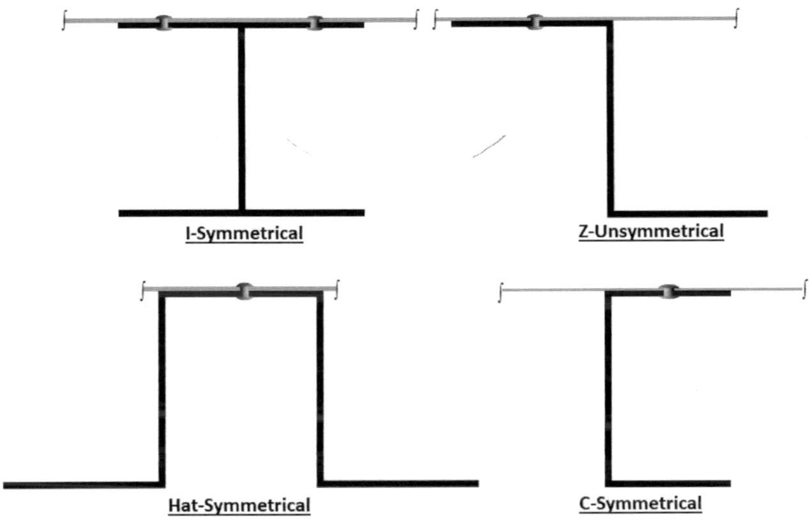

Fig. 5: Typical stringer configurations.

Title: Progressive Damage and Failure Analysis for Structural Continuous Fiber Composites

Abstract:

Composite materials, widely used in aircraft components and sporting equipment, are renowned for their exceptional strength-to-weight ratios. However, their inherent anisotropy and the combination of different constituents (fibers and matrix) make them prone to various failure modes, including matrix breakage, fiber failure, and delamination.

In the aerospace and defense (A&D) sector, evaluating damage in composite structures is a significant challenge. The certification process requires rigorous numerical model verification and validation (V&V), which consumes substantial resources. Additionally, the steep learning curve associated with these processes further complicates the task.

To address these challenges, a new simulation workflow for progressive damage modeling in composite materials has been developed. This workflow leverages the Building Blocks Approach, which breaks down the design and manufacturing process into discrete components or modules. Similar to assembling building blocks to construct a larger structure, this method allows for greater flexibility, efficiency, and optimization throughout the entire lifecycle of the composite component.

The proposed workflow focuses on efficient preprocessing, standardized procedures, and enhanced collaboration with industry partners. By streamlining the creation of high-fidelity simulation input decks, this approach aims to reduce the time and effort required for V&V, ultimately accelerating the certification process. This innovative workflow not only simplifies the simulation process but also enhances the accuracy and reliability of damage assessments in composite materials.

Keywords: Composite materials, failure modes, matrix breakage, fiber failure, delamination, aerospace and defense (A&D), damage assessment, building blocks approach, structural carbon fiber reinforced plastic (CFRP), simulation workflow

MULTIFUNCTION AUTOMATED UNIT CELL REVIEW

S.Shoham[*1], U.Ben-Simon[1], Z. Karuchero[1], A.Lukatsky[1], A.Sawday[1], E.Ejgenberg[1] , Y. Freed[1] & H.Elimelech[2]

[1]Israel Aerospace Industries, Ben-Gurion International Airport, Israel
[21]DDR&D- Directorate of Defense Research & Development, Tel-Aviv, Israel

INTRODUCTION

Automated layup fabrication has transformed the landscape of aerostructure manufacturing over the past few decades, driven by a convergence of critical factors: (1) sustainability regulations and evolving public opinion on environmental issues and technological advancements, (2) the imperative of lifecycle cost reduction as a key determinant of customer satisfaction, (3) the developed markets' need to maintain competitiveness against low-labor-cost economies, and (4) advancements in technology enabling real-time monitoring and analysis of multiple key parameters. This review explores the integration of Multifunction Fabrication and Curing Processes (MFCP) using automated cells within the aviation aerostructure industry. A comprehensive overview of current automated fabrication and curing processes is presented, the MFCP cell concept is outlined, and the driving forces behind its potential adoption are analyzed.

FROM STATE OF THE ART TO NEXT GENERATION

The paper begins with a review of state-of-the-art technologies for fabrication of composite materials, including Automated Tape Layup (ATL), Automated Fiber Placement (AFP), Resin Transfer Molding (RTM), Liquid Resin Injection (LRI), and additive manufacturing methods such as Powder Bed Fusion and Fused Deposition Modeling (FDM). Case studies illustrating their application in aerostructure manufacturing are provided, with special emphasis on integration of several technologies in a single fabrication cell. Additionally, emerging thermoplastic composite technologies, including Organo Sheets, Over Molding Injection (OMI), Sheet Molding Compound (SMC), and Bulk Molding Compound (BMC), are examined for their potential integration into aerostructure production.

SUMMARY

The review is concluded by evaluating the adoption challenges and opportunities of the MFCP concept within the aviation sector. Early adoption is anticipated within military and defense applications due to their autonomy to manage stringent certification requirements, which pose significant challenges for civil applications. Current limitations in element sizing also hinder adoption in regional and business jet segments major structural elements. Further enhancements, such as automated non-destructive testing (NDT), cleaning, and trimming processes, are proposed to maximize the efficiency of a single-stop manufacturing approach. The analysis suggests that the adoption of MFCP will depend on a business environment prioritizing agility, flexibility, and lead time over cost-effectiveness and stringent quality certification, making defense applications the most probable early adopters of this innovative concept.

Manuscript 26 submitted to 64th Israel Annual Conference on Aerospace Sciences (IACAS). Received December 24, 2024.

Optimal design of thin-walled beams with buckling considerations

Ameer Marzok

Department of Aerospace Engineering, Technion – Israel Institute of Technology, Haifa, 3200003

Thin-walled beams have many applications in various engineering disciplines including aerospace, mechanical, and civil engineering. While these elements have slender cross-sections that lead to a desired high stiffness-to-volume ratio, they are prone to global and local buckling. Therefore, buckling considerations are essential to ensure the adequate performance of these elements.

This contribution presents a comparison between two different approaches, developed by the author and his colleges, for the design of these elements utilizing structural optimization techniques. The first method, referred to as TopOpt, is based on topology optimization of the cross-section, and the second, referred to as GrOpt, relies on the ground structure approach.

In TopOpt, the cross-section is discretized into density variables, providing the possibility to attain general shapes in a continuous distribution of the material within the domain. This model relies on an efficient topology optimization procedure, developed for this problem, that uses a novel extended/generalized finite element method (XFEM/GFEM) with global enrichment functions to analyze the displacements and buckling loads of these elements. The method alleviates the need for detailed 3D finite element models, even when complex deformation modes are involved.

GrOpt considers design variables that determine the location and thickness of line segments, which form the beam's cross-section. The structural analysis in this model relies on an efficient semi-analytical finite strip model (FSM).

The optimization is conducted using a gradient-based optimization algorithm with analytical derivation of the sensitivity analysis. It is noted that the computational burden of the methods is significantly different. However, the final designs obtained are qualitatively similar.

Several case studies will be simulated using both methods, aiming to minimize the weight of the beams while accounting for displacement and buckling constraints. The results will be presented, illustrating the similarities and differences between the methods.

Manuscript 22 submitted to 64th Israel Annual Conference on Aerospace Sciences (IACAS). Received December 23, 2024.